Climate Change
Essential Guide to Science, Politics and Solutions

Other World Scientific Titles on Climate Change

Waking Up to Climate Change: Five Dimensions of the Crisis and What We Can Do About It
by George Ropes
ISBN: 978-981-12-4623-4
ISBN: 978-981-12-4754-5 (pbk)

Science of the Earth, Climate and Energy
by Milton W Cole, Angela D Lueking and David L Goodstein
ISBN: 978-981-323-361-4
ISBN: 978-981-327-178-4 (pbk)

How to Cope with Climate Change
by Michael Krause
ISBN: 978-981-12-8683-4
ISBN: 978-981-12-8739-8 (pbk)

The Ninth Revolution: Transforming Food Systems for Good
by Sayed Nader Azam-Ali
ISBN: 978-981-12-3644-0
ISBN: 978-981-12-5010-1 (pbk)

Electricity: Humanity's Low-carbon Future: Safeguarding Our Ecological Niche
by Hans B (Teddy) Püttgen and Yves Bamberger
ISBN: 978-981-12-2435-5
ISBN: 978-981-12-2930-5 (pbk)

Climate Change and Energy Options for a Sustainable Future
by Dinesh Kumar Srivastava and V S Ramamurthy
ISBN: 978-981-12-3347-0
ISBN: 978-0-00-099006-8 (pbk)

Climate Change
Essential Guide to Science, Politics and Solutions

Mark Maslin
University College London, UK
United Nations University, Canada

World Scientific
NEW JERSEY • LONDON • SINGAPORE • BEIJING • SHANGHAI • HONG KONG • TAIPEI • CHENNAI • TOKYO

Published by

World Scientific Publishing Co. Pte. Ltd.
5 Toh Tuck Link, Singapore 596224
USA office: 27 Warren Street, Suite 401-402, Hackensack, NJ 07601
UK office: 57 Shelton Street, Covent Garden, London WC2H 9HE

Library of Congress Control Number: 2026938774

British Library Cataloguing-in-Publication Data
A catalogue record for this book is available from the British Library.

Warming Stripes image on cover courtesy of Professor Ed Hawkins

CLIMATE CHANGE
Essential Guide to Science, Politics and Solutions

ISBN 978-981-98-2649-0 (hardcover)
ISBN 978-981-98-2718-3 (paperback)
ISBN 978-981-98-2650-6 (ebook for institutions)
ISBN 978-981-98-2651-3 (ebook for individuals)

For any available supplementary material, please visit
https://www.worldscientific.com/worldscibooks/10.1142/14682#t=suppl

Desk Editors: Zerlina Zhuang/Amanda Yun

Typeset by Stallion Press
Email: enquiries@stallionpress.com

Dedication

To my father, who showed me that humble beginnings should never limit your aspirations, your dreams, or your achievements.

Acknowledgements

The author would like to thank the following people: Shooka Bidarian for risking everything to build an amazing life with me and for encouraging me to write another book on climate change. Chris Maslin, my father, for always supporting me from picking me up at 3 am as a teenager after being stranded in the countryside after a rock concert to buying me my first computer to write my PhD on, and much more. Alexandra and Abbie Maslin for being amazing daughters. Frankie Alexander my first grand-child for being so cute and of course if he reads this when he is older he will hate me. Professor Lisa Vanhala who it has been an honour to job share the role of Pro-Vice Provost of the UCL Climate Crisis Grand Challenge. All the amazing colleagues at UCL, who are too numerous to name, who I have worked with, written papers with and taught with (you know who you are — thank you!). And last, but not least, Miles Irving for excellent illustrations for all my books.

About the Author

Mark Maslin BSc, PhD, FRGS, FRSA is a Professor of Earth System Science at University College London (UCL), UK and the Lead on Climate, Health and Security at United Nations University (UNU-INWEH), Canada. He is the Pro-Vice Provost of the UCL Climate Crisis Grand Challenge. Professor Maslin is a leading scientist with particular interest in understanding climate change in the past, present and future. His work focuses on understanding the Anthropocene and the major challenges facing humanity in the 21st century.

Professor Maslin has published over 230 papers in journals such as *Science, Nature, and The Lancet*. He has received research, consultancy and training funding worth over £85m from government, charities, NGOs and the private sector. He has written 10 books and over 100 popular articles and regularly appears on radio and television. These appearances include BBC One David Attenborough's 'Climate Change — The Facts', as well as 'Dispatches', 'Newsnight', 'Time Team', 'The Today programme', 'In Our Time', 'Inside Science' and 'The Briefing Room'. His books include *The Human Planet* with Prof. Simon Lewis (2018), *Climate Change: A Very Short Introduction* (2021), *The Cradle of Humanity* (2019), *How to Save Our Planet: The Facts* (2021) and *Climate: A Very Short Introduction* (2026).

Professor Maslin was included in *Who's Who* for the first time in 2009. He is a Royal Society Industrial Fellow and a Royal Society Wolfson Research Merit Scholar. He is also the United Nations Framework Convention on Climate Change (UNFCCC) designated point of contact and Conference of the Parties (COP) meeting lead for UCL. In 2023, *Onalytical* listed Professor Maslin as the top Global Sustainability thought leader and influencer of the year. In 2024, he was the Special Advisor to the UK Parliament Environment Audits Committee.

Contents

Abbreviations

AABW	Antarctic Bottom Water
AO	Arctic Oscillation
AOGCM	Atmosphere–Ocean General Circulation Model
AOSIS	Alliance of Small Island States
CCS	Carbon capture and storage
CO_2	Carbon Dioxide
CDM	Clean Development Mechanism
CFCs	Chlorofluorocarbons
CH_4	Methane
CMIP6	Coupled Model Intercomparison Projects
COP	Conference of the Parties
CORSIA	Carbon Offsetting and Reductio Scheme for International Aviation
ECS	Equilibrium Climate Sensitivity
ENSO	El Niño–Southern Oscillation
ETS	Emissions Trading Scheme
EV	Electric vehicle
GCM	General circulation model
GCR	Galactic cosmic ray
GDP	Gross Domestic Product
GHCN	Global Historical Climate Network
GHG	Greenhouse gas
GMT	Global mean temperature
GO	Guarantee of Origin

GtC	Gigatonnes of Carbon
H_2O	Water vapour
IEA	International Energy Agency
IFRS	International Financial Reporting Standards
IMF	International Monetary Fund
IPCC	Intergovernmental Panel on Climate Change
I-REC	International Renewable Energy Certificates
ITCZ	Intertropical Convergence Zone
MAT	Marine air temperature
MRV	Monitoring, Reporting, and Verification
N	Nitrogen
NADW	North Atlantic Deep Water
NAO	North Atlantic Oscillation
NASA	National Aeronautics and Space Administration
NDCs	National Determined Contributions
NGO	Non-governmental organisation
NOAA	National Oceanic and Atmospheric Administration
NRC	National Research Council (USA)
N_2O	Nitrous Oxide
O_2	Oxygen
O_3	Ozone
OECD	Organization for Economic Cooperation and Development
OPEC	Organization of the Petroleum Exporting Countries
PDO	Pacific Decadal Oscillation
PETM	Palaeocene–Eocene Thermal Maximum
ppbv	Parts per billion by volume
ppmv	Parts per million by volume
RCP	Representative Concentration Pathways
REC	Renewable Energy Certificates
REDD	Reduced Emissions from Deforestation and Forest Degradation
SBTi	Science Based Targets initiative
SRES	Special Report on Emission Scenarios by the IPCC (2000)
SSP	Shared Socioeconomic Pathway
SSS	Sea-surface salinity
SST	Sea-surface temperature
TCFD	Task Force on Climate-related Financial Disclosures

THC	Thermohaline circulation
UBI	Universal Basic Income
UN	United Nations
UNCED	United Nations Conference on Environment and Development
UNDP	United Nations Development Programme
UNFCCC	United Nations Framework Convention on Climate Change
UV	Ultravoilet
VBD	Vector-borne disease
VCMs	Voluntary carbon markets
WTO	World Trade Organization
WSO	World Sustainability Organization
WHO	World Health Organization
WEO	World Environment Organization

List of Illustrations

Chapter 1

What is Climate?

Introduction

Climate affects everything we do in life; from the clothes we wear to the diseases we catch. As humans, we only feel comfortable within a very narrow range of temperature and humidity. This comfort zone ranges from about 20°C to 26°C and from 20% to 75% relative humidity (see Figure 1)[1]. We live across the whole globe in conditions that are frequently outside this comfort zone. We have learnt to adapt our clothing and dwellings to maintain our comfort. So, while you may think the clothes you have hanging in your wardrobe simply reflect your fashion taste or lack of, in reality the clothes reflect the climate in which you live and how it changes throughout the year. For example, you may have a thick padded coat for a Canadian winter and a short-sleeved shirt for a business meeting in Jakarta. Our wardrobes also give hints about where we like to take our holidays. If you are a budding Polar explorer, then there will be very warm Arctic clothes hanging up; if you love sunning yourself on the beach, then there will be shorts or a bikini instead.

Our houses are also built with a clear understanding of local climate. In England, almost all houses have central heating as the outside temperature is usually below 20°C, but very few have air conditioning as temperatures rarely exceed 26°C. On the other hand, in Singapore, most houses have air conditioning but rarely central heating. Climate also affects the structure of our cities and how transport systems around the world operate.

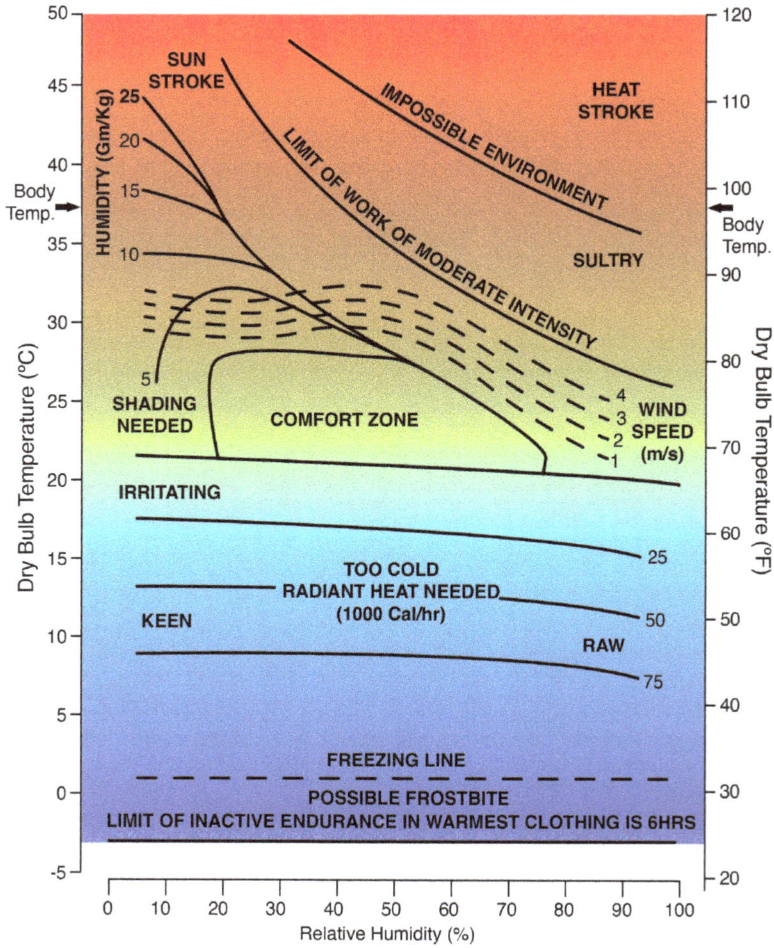

Figure 1. Human Comfort and Climate[1]

In Houston, Texas, there is a network of 7 miles of underground tunnels connecting all the major downtown buildings; this is fully climate controlled and links 95 heavily populated city blocks. People use the tunnel when it is raining or hot outside, because for at least 5 months of the year, the average temperature in Houston is above 30°C. Similarly, there are underground malls in Canada but in this case, it is to avoid the problems of heavy snow and extreme cold.

Climate controls where we grow our food and when certain food is available. This is because agriculture is controlled by temperature, rainfall, the occurrence of frost and snow, and by the length of the growing season. Simplified examples are: rice is grown where it is warm and very wet, while wheat can grow in much more temperate climes. The climate can also affect the quality of our food, for example: it is well known that the very best vintages of French wine are produced when there are a few short sharp frosts during the winter which harden the vines, hence, producing excellent grapes. Farmers can also 'help' the local climate, for example by growing tomatoes in a greenhouse or by irrigating the land to provide a more constant supply of water.

Climate also influences where there will be extreme weather events such as heatwaves, droughts, wildfires, floods, and storms. However, in many cases, our perception of extreme events is determined by local conditions, so for example: in 2022, England experienced a 'heatwave' and 40°C was recorded for the first time ever in the UK[2]. In many countries in the tropics, a heatwave would not be recorded until temperatures were well above 45°C. Climate also has a large effect on our health, as many diseases are temperature and humidity controlled. For example: incidences of influenza, commonly called the flu, reaches a peak in winter. Since the Northern and Southern Hemispheres have winter at different times of the year, there are actually two different flu seasons globally each year. The influenza virus migrates between the two hemispheres after each winter, giving us time to produce new vaccinations based on the new strain of flu that has appeared in the previous six months in the other hemisphere.

Hot and Cold Earth

The climate of our planet is caused by the Equator of the Earth receiving more of the Sun's energy than the poles. If you imagine the Earth is a giant ball, the closest point to the Sun is the middle or the Equator. The Equator is where the Sun is most often directly overhead, and it is here that the Earth receives the most energy. As you move further north or south away from the Equator, the surface of the Earth curves away from the Sun, increasing the angle of the surface of the Earth relative to the Sun. This means the Sun's energy is spread over a larger area causing less warming. If we lived on a flat disc, we would get much more energy from the Sun — about 1,370 Watts per square metre (W/m^2) — instead the planet surface averages about 343 W/m^2 (Figure 2) due to its curved nature[3]. The Earth also receives a very small fraction of the energy pumped out of the Sun. If you consider how small the Earth is compared with the Sun, for every Watt we receive from the Sun, it emits 2 billion Watts. This is why in many science fiction novels, the authors imagine a strip or even a sphere around a star to collect all that energy that is simply being lost into space.

About one-third of the solar energy we receive is reflected straight back into space. This is because of 'albedo', which is reflectivity of a surface. For example: white clouds and snow have a very high albedo and reflect almost all of the sunlight that falls on them, while darker surfaces such as the oceans, grassland, and rainforest absorb a lot more energy. Not only do the poles receive less energy than the Equator, but they also lose more energy back into space (Figure 3): the white snow and ice in the Arctic and Antarctic have a high albedo and bounce a lot of the Sun energy back into space. On the other hand, the darker, much less reflective

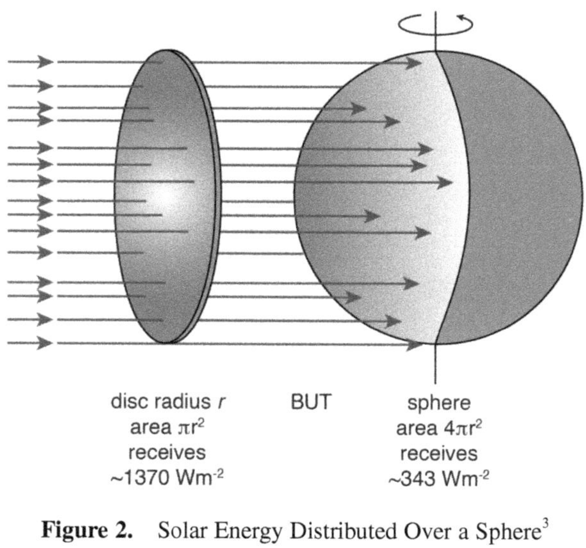

Figure 2. Solar Energy Distributed Over a Sphere[3]

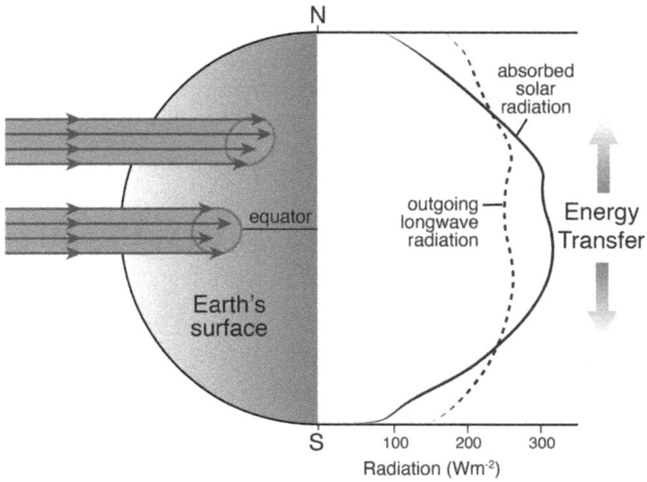

Figure 3. Energy Transfer Away from the Equator Driven by Angle of Solar Radiation[3]

vegetation at lower latitudes absorbs a lot more energy. These two processes working together mean that the tropics are hot and the poles are very cold. Nature hates this sort of energy imbalance, so energy, in the form of heat, is transported by both the atmosphere and oceans from the Equator to both poles, and this affects the climate.

Seasons on Earth

Our climate is controlled by two fundamental facts about the relationship between the Earth and the Sun. The first is the tilt of the Earth's axis of rotation, which causes the seasons. The second factor is the daily rotation of the Earth that provides us with night and day, as well as drives the circulation of both the atmosphere and the oceans.

The Earth's axis of rotation is tilted at an angle of 23.5° and results in a seasonal difference in the amount of energy received by each hemisphere throughout the year. The seasonal changes are by far the largest effect on climate. It is amazing to think that if the Earth were not tilted and stood straight up on its axis, then we would not have spring, summer, autumn, and winter. We would not have the massive change in vegetation in the temperate latitudes, and we would not have the monsoon and hurricane seasons in the tropics. The reason for the seasons is the change in the angle of the sunlight hitting the Earth through the year. If we take 21 December as an example, the Earth's axis is leaning away from the Sun, so the sunlight hitting the Northern Hemisphere is at a greater angle and spreading its energy over a wider area creating Northern Hemisphere winter. In the Arctic the lean is so great that sunlight cannot even reach the surface and this produces 24 hours of darkness. Everything is opposite in the Southern Hemisphere, since it is then leaning towards the Sun and hence the sunlight is more directly overhead. This means that Antarctica is bathed in 24 hours of sunlight and people in Australia have Christmas dinner on the beach, while topping up their tan. As the Earth moves round the Sun, taking about 365.25 days (hence the leap year every fourth year), the angle of the axis stays in the same place. Hence when it comes to June, the Earth's axis is leaning towards the Sun, so the Northern Hemisphere has lots of direct sunlight and thus it is summer, while the Southern Hemisphere is shielded from the sunlight and is plunged into winter.

If we follow the Sun through a single year, we can see how this tilt affects the Earth through the seasons. If we start at 21 June, the Sun is overhead at midday at the Tropic of Cancer (23.4°N), the northern summer solstice. The angle of the Sun moves southward until 21 September when it is overhead at midday over the Equator, the equinox or autumn equinox in the Northern Hemisphere. The Sun appears to continue southward and on 21 December, it is overhead at midday at the Tropic of Capricorn (23.4°S) the southern summer solstice. The Sun then appears to move northward until it is directly overhead at midday at the Equator on 21 March, the equinox or spring equinox in the Northern Hemisphere and so the cycle continues (see Figure 4).

The seasons signal by far the most dramatic change in our climate; if we take for example Seoul (South Korea), winter temperatures can be as low as −20°C while summer temperatures can be over 38°C, that is a 58°C temperature difference. The seasons are also one of the major reasons for storms — leading to hurricanes, cyclones, typhoons, tornadoes, monsoons and winter storms[4].

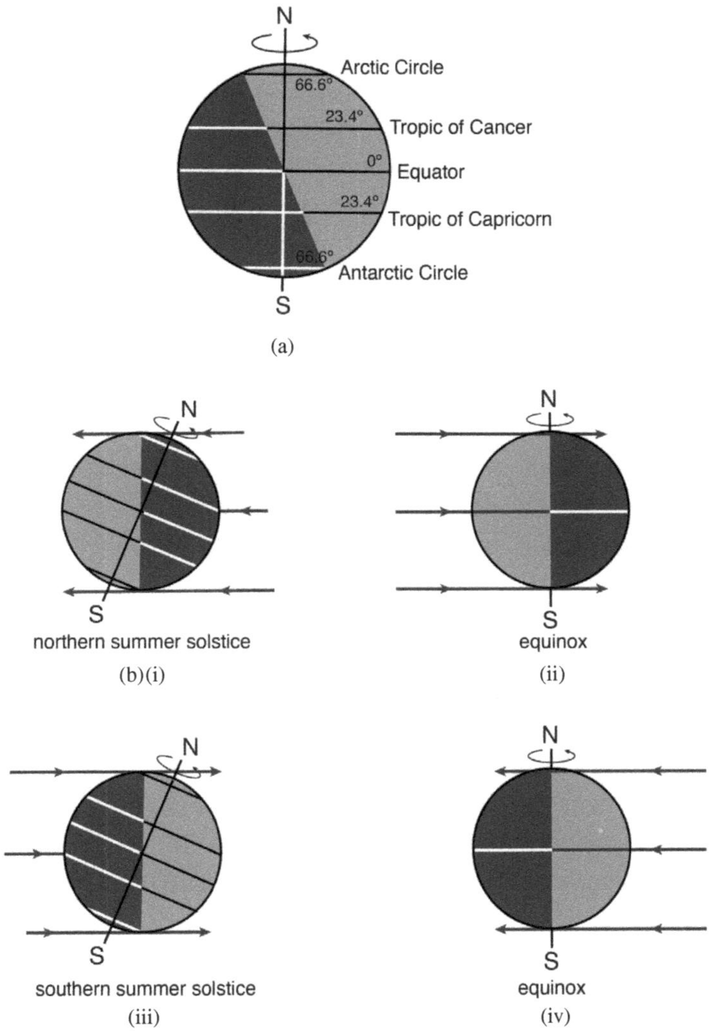

Figure 4. Solstice and Equinox Caused by the Tilt of the Earth[3]

Moving Heat Around the Earth

The second big factor affecting the climate of the Earth is its daily rotation. First this plunges the Earth in and out of darkness causing massive changes in diurnal temperature. For example, in the Gobi Desert during summer daytime, temperatures can be over 38°C and then nighttime lows of just 5°C; while Hong Kong has a diurnal temperature range of little more than 4°C. Depending on the season, different areas also get varying amounts of daylight. The days can vary from 24 hours' daylight to 24 hours' darkness at the poles to around 12 hours' sunlight every day at the Equator. This change in the daylight compounds the seasonal contrasts, because not only during summer do you get more direct 'overhead' sunlight, but also you get it for much longer.

The spinning of the Earth also makes the transport of heat away from the Equator more complicated. This is because the spinning of the Earth makes everything else, including the atmosphere and oceans, turn. The simple rule is that rotation of the Earth causes the air and ocean currents to be pushed to the right of the direction they are travelling in the Northern Hemisphere and to the left of the direction they are travelling in the Southern Hemisphere. This deflection is called the Coriolis Effect and its strength increases the further you go towards the poles[4].

An everyday example of this, which is always quoted, is the way water flows down a plughole or a toilet. In the Northern Hemisphere water is said to flow clockwise down the plughole while in the Southern Hemisphere it is anti-clockwise. However, I hate to tell you that the direction the water drains out of your bath or toilet is not related to the Coriolis effect or to the rotation of the Earth. There is no observable consistent difference in rotation direction between toilets in the Northern and Southern Hemispheres. This is because the Coriolis Effect has such a small influence compared with any residual movement of the water and the effect of the shape of the container. This also mean the wonderful cottage industry of communities living on the Equator showing tourists the Coriolis effect is simply done by a sleight of hand. For example: in Kenya, there are big signs telling you when you are crossing the Equator; if you care to stop at the road side, locals will happily pour water from a bucket into a large funnel and seeming to demonstrate clearly that it goes a different way round when you are standing on one side of the sign than when you are standing on the other. However, this change is all in the wrist and how the water is poured in, affecting which way it goes round. Still, even though it is completely fake, I love these demonstrations as it means loads of locals and tourists get to hear about the Coriolis Effect.

Coming back to climate, why then do the ocean currents and winds have this deflection? Imagine firing a missile from the Equator directly north. Because the missile was fired from the Earth, which is spinning eastward, the missile is also moving east. As the Earth spins, the Equator has to move fast through space to keep up with the rest, as it is the widest part of the Earth. As you go further north or south away from the Equator, the surface of the Earth curves in, so it does not have to move as fast to keep up with the Equator. So, in one day, the Equator must move round 40,074 km (the diameter of the Earth) with a speed of 1,670 km/hour, while the Tropic of Cancer (23.4°N) has to move 36,750 km, with a speed of 1,530 km/hour, and the Arctic circle (66.6°N) has to move 17,662 km at a speed of 736 km/hour. At the North Pole, there is no relative movement at all so the speed there is 0 km/hour. A practical demonstration of this is if you hold hands with a friend and stand in the same place while spinning them around, they will travel much faster than you do. The missile, fired from the Equator, has the faster eastward speed of the Equator; as it moves northward towards the Tropic of Cancer, the surface of the Earth is not moving as fast eastward as the missile. This gives the appearance that the missile is moving northeast as it is moving faster eastward than the area it is moving into. As you get closer to the poles, the greater this difference in speeds so the greater the deflection to the east (see Figure 5).

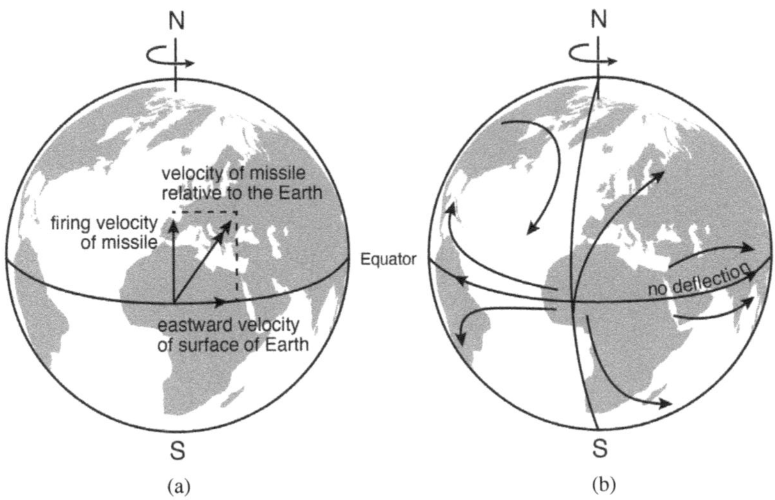

Figure 5. Coriolis Effect Due to Relative Movement of the Surface of the Earth

The Atmosphere and Oceans

The atmosphere is the home of our weather. It begins at the surface of the Earth and becomes thinner and thinner with increasing altitude, with no definite boundary between the atmosphere and outer space. The arbitrary Kármán line at 62 miles (100 km), named after Theodore von Kármán (1881–1963), a Hungarian-American engineer and physicist, is usually used to mark the boundary between atmosphere and outer space. The layer of atmosphere in which weather takes place is thinner at about 10 miles thick. The oceans also play an important part in controlling our weather and climate. The oceans cover 70% of the Earth's surface and are on average about 2.5 miles deep, so the total thickness of the layer controlling our climate is 12.5 miles thick[4].

The atmosphere is a mechanical mixture of gases, not a chemical compound. What is significant is that these gases are mixed in remarkably constant proportions up to about 50 miles (80 km) above the surface of the Earth. Four gases — nitrogen, oxygen, argon, and carbon dioxide — account for 99.98% of air by volume. Of special interest are the greenhouse gases that despite their relative scarcity, have a great effect on the thermal properties of the atmosphere, which include carbon dioxide, methane, and water vapour. These are discussed in the next chapter.

Content of the Atmosphere

Nitrogen is a colourless, odourless, tasteless, and mostly inert gas and makes up ~78% by volume of the Earth's atmosphere[5]. Argon is also a colourless, odourless, tasteless, and completely inert gas and makes up ~0.9% by volume of the Earth's atmosphere. In contrast, oxygen is a very reactive gas and makes up ~21% of the Earth's atmosphere by volume. Oxygen sustains all life on Earth and is constantly recycled between the atmosphere and the biological processes of plants and animals. It combines with hydrogen to produce water, which in its gaseous state, water vapour, is one of the most important components of the atmosphere as far as weather is concerned.

Oxygen also forms another gas called ozone or trioxygen, which is made up of three oxygen atoms instead of the usual two. This is an extremely important gas in the atmosphere as it forms a thin layer in the stratosphere (between 6 and 31 miles) that filters out harmful ultraviolet radiation that can cause cancer. Even in this 'layer', the ozone concentrations are only two to eight parts per million in volume, so most of the oxygen remains of the normal dioxygen type. Much of this important gas was being destroyed by our use of CFCs, and ozone holes have been found over the Arctic and Antarctic, until governments worldwide agreed (for example, in the Vienna Convention for the Protection of the Ozone Layer in 1985 and then in the Montreal Protocol on Substances that Deplete the Ozone Layer in 1987) to stop the use of all CFCs and related compounds.

Carbon dioxide makes up 0.04% of the Earth's atmosphere and is a major greenhouse gas, important for keeping the Earth relatively warm[5]. Until recently, the level of carbon dioxide has been balanced through its consumption by plants for photosynthesis and its production by plants and animals in respiration. Over the last 150 years, human industry and land use changes have emitted huge amounts of carbon dioxide into the atmosphere, upsetting this natural balance.

Aerosols are suspended particles of sea salt, dust (particularly from desert regions), organic matter, and smoke. The height at which these aerosols are introduced will determine whether they cause regional warming or regional cooling. This is because high up in the atmosphere they help reflect sunlight, thus cooling the local area, while at low altitudes, they absorb some of the warmth coming off the Earth, thus warming the local air. Industrial processes have increased the level of aerosols in the atmosphere, which has led to smog in urban areas, acid rain, and localised cooling. But the most important effect of aerosols is to help clouds form. Without these minute particles, water vapour cannot condense and form clouds; and without cloud precipitation, there is no weather.

Water vapour is the forgotten but most important greenhouse gas, which makes up about 1% by volume of the atmosphere, but is highly variable in time and space as it is tied to the complex global hydrological cycle[5]. The most important role that water vapour plays in the atmosphere is the formation of clouds and the production of precipitation (rain or snow). Warm air can hold more water vapour than cold air. When a parcel of air cools down, for example as the air rises or meets a cold air mass, it cannot hold as much water vapour, so the water condenses on to aerosols and produces clouds. An important point which we will discuss later is that as water changes from a gas to a liquid, it releases some energy, and it is this energy which can fuel storms as large as hurricanes. Clouds come in all sorts of shapes and sizes and are an excellent way of telling what sort of weather is coming soon.

Hadley, Ferrel, and Polar Cells

The shape of the Earth creates a temperature imbalance between the Equator and the poles. Both the atmosphere and the oceans act as transporters for this heat away from the Equator. But as always with climate, things get a little more complicated. At the Equator, the intense heat from the Sun warms up the air near the surface and causes it to rise high into the atmosphere. Warm air rises because the gas molecules in warm air can move further apart making the air less dense, and correspondingly, cold air sinks. This rising air creates a space and lowers atmospheric pressure, which is filled by air being sucked in laterally. This produces the Trade Winds in both the North and South Hemispheres. The northeast and southeast Trade Winds meet at the Inter-Tropical Convergence Zone (ITCZ). This causes a problem as the climate system is desperately trying to export heat away from the region around the Equator, and these in-blowing winds do nothing to help this removal of heat. In the tropics, it is the surface currents of the ocean that transport most of the heat (see Figure 6)[1]. These currents include the Gulf Stream, which takes heat from the tropical Atlantic and transports it northward keeping Europe's weather mild all year round (see Box 2). Other currents include the Kuroshiro current in the western North Pacific, the Brazilian current in the western South Atlantic, and finally the East Australian current in the western South Pacific.

The tropical hot air which has risen high into the atmosphere slowly cools, due to both its rise and its movement towards the poles, and at about 30° north and south it sinks, forming the sub-tropical high-pressure zone. As this sinking air reaches the surface, it spreads out, moving both north and south. This sinking air has lost most of its moisture and therefore dries

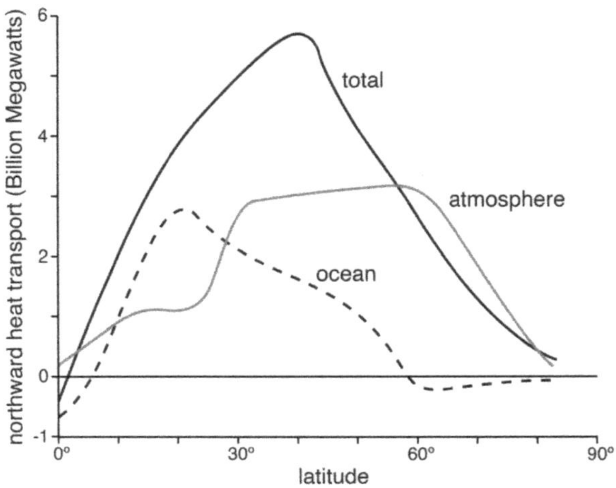

Figure 6. Heat Transport Away from the Equator

out the land it sinks onto, producing some of the major deserts around the world. The southward air links into the first atmospheric cell called the Hadley Cell and becomes part of the Trade Wind system. The northward-bound air forms the Westerlies and it is from here northwards that the atmosphere takes over from the oceans as the major transporter of heat. The movement of warm sub-tropical air northward is only stopped when it meets the cold Polar air mass at the Polar Front. The intense cold at the poles causes air to become super chilled and sink, causing out-blowing winds (Figure 7). When this cold Polar air meets the warm, moist Westerlies at the Polar Front, the clash causes the Westerlies to lose a lot of their moisture in the form of rain. It also forces the warm sub-tropical air to rise, as the cold Polar air is much heavier. This rising air completes the other two cells, the Ferrel or mid-latitude cell, and the Polar Cell — because as the air rises, it spreads out to both the north and south. To the south, this high-rise air meets with tropical air coming northward and sinks forming the middle Ferrel Cell. The northward component of this rising air drifts over the poles where it is chilled and sinks, forming those Polar out-blowing winds which complete the third Polar Cell (Figure 7). The names of two of the three cells come from George Hadley, an English lawyer and amateur meteorologist, who in the early 18th century explained

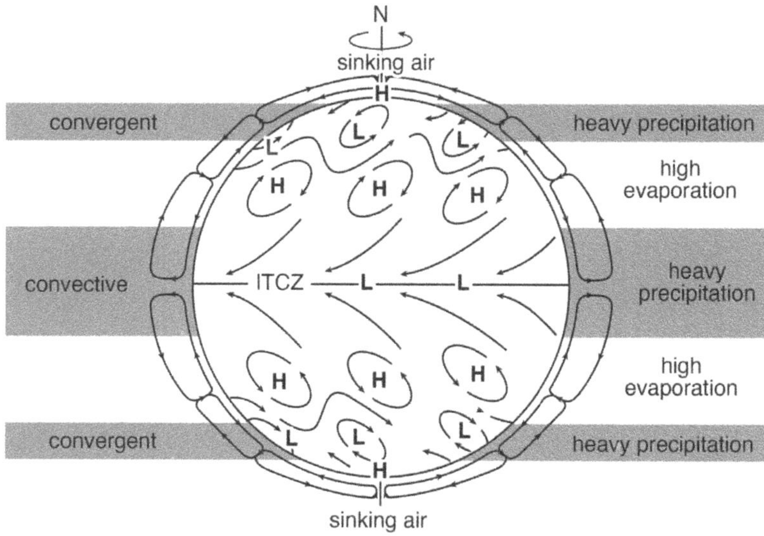

Figure 7. Major Atmospheric Circulation Patterns

the mechanism which sustained the Trade Winds. In the mid-19th century, William Ferrel, an American meteorologist, developed Hadley's theories by explaining the mid-latitude atmospheric circulation cell in detail.

An important component of these cells is the high altitude, fast flowing, narrow air currents called jet streams[6]. The main jet streams are located near the tropopause, which represents the transition between the troposphere and the stratosphere. The major jet streams are westerly winds that flow west to east. Their paths typically have a meandering shape; jet streams may start, stop, split into two or more parts, combine into one stream, or flow in various directions including the opposite direction of most of the jet. The strongest jet streams are the polar jets, at around 7–12 km above sea level, and the higher and somewhat weaker sub-tropical jets at around 10–16 km (see Figure 8). The Northern and the Southern Hemispheres each have both a Polar jet and a sub-tropical jet. The Northern Hemisphere Polar jet flows over the middle to northern latitudes of North America, Europe, and Asia and their intervening oceans, while the Southern Hemisphere Polar jet mostly circles Antarctica all year round. Jet streams are caused by a combination of the Earth's rotation and

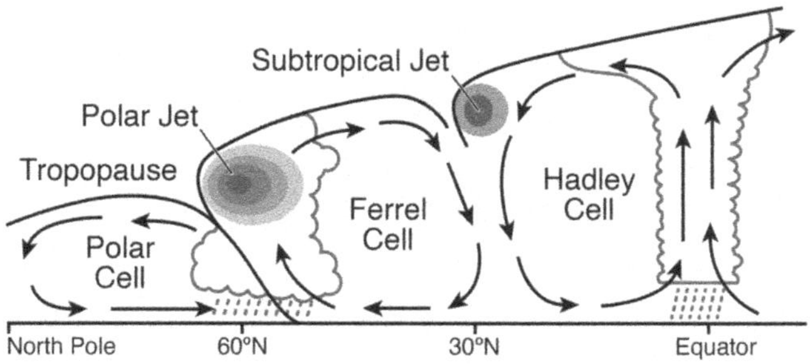

Figure 8. Major Atmospheric Circulation Cells and Jet Streams

energy in the atmosphere; hence they form near boundaries of air masses with significant differences in temperature (Figure 8).

Though the general wind patterns of Earth follow this simple three-celled, two jet stream per hemisphere model, in reality they are much more complicated. Consider the first reason: the Earth is spinning and it adds the influence of the Coriolis Effect on the air. This means that air masses trying to flow northward or southward are deflected by the spinning of the Earth. For example, this causes large meanders in the jet streams, which are called planetary waves. These can have a huge effect on our weather. Unusual deflections of these planetary waves cause disasters such as the extreme flooding in Japan in 2018, China in 2021 and South Korea in 2025. Besides flooding, there have also been extreme heatwaves, including China in 2025, when the township in Xinjiang, in northwestern China, hit 52.5°C — the highest recorded temperature in China. In Siberia, there was a record-breaking 38°C in the town of Verkhoyansk on 20 June 2020, causing massive wildfires, loss of permafrost, and an invasion of pests. The opposite — extreme cold — happens as well. Examples included the big freeze of Beijing with nearly two weeks of temperatures below –10°C and the Texas cold snap of 2021 with temperatures down to –19°C.

The second complication is that the continents heat up much quicker than the oceans, which can cause the surface air over the land to rise, potentially altering the general circulation of surface wind. This then leads to local land–sea breezes and, on a much larger scale, creating the monsoon systems. The seasons, then, can have a huge effect on atmospheric circulation, because during the summer in each hemisphere, the land heats up much more than the ocean, causing the ITCZ to be pulled southwards towards Australasia, and across South America and Southeast Africa during Southern Hemisphere summer and northwards across India, Southeast Asia and North Africa during Northern Hemisphere summer.

The Hadley Cells explain why there are three main rainfall belts across the Earth, the convection rainfall belt which moves north and south of the Equator and the two convergent rainfall belts, one in the Northern and one in the Southern Hemisphere where warm, moist subtropical air meets cold dry Polar air. They also explain why there are two main desert belts in the world, which are usually found between the rainfall belts with super dry air sinking between the Hadley and Ferrel Cells. In the Northern Hemisphere, good examples are the Sahara Desert in North Africa and the Gobi Desert in China, while in the Southern Hemisphere, Central Australia Desert and the Kalahari Desert in South Africa are good examples.

The Hadley Cells can also be used to define the three main storm zones. First are 'winter storms' at the Polar Front. Second are the subtropical highs and the Trade Wind belt, which are the spawning ground for hurricanes. Third is the ITCZ, where the rapidly rising air cools and produces tropical thunderstorms with heavy rainfall, producing monsoons as it moves over the land.

Ocean Circulation

As we have seen, the surface ocean is important in transporting heat around the globe. The circulation of the oceans starts with the wind, because it is the action of the wind on the surface ocean that makes it move (Figure 9)[4]. As the wind blows on the surface water, the friction allows energy to be transferred from the winds to the surface water, leading to major currents. The wind energy is transferred to greater depths in the water column turbulence, which allows wind driven currents to be very deep.

The circulation of the deep ocean is one of the major controls on global climate due to its ability to exchange heat between the two

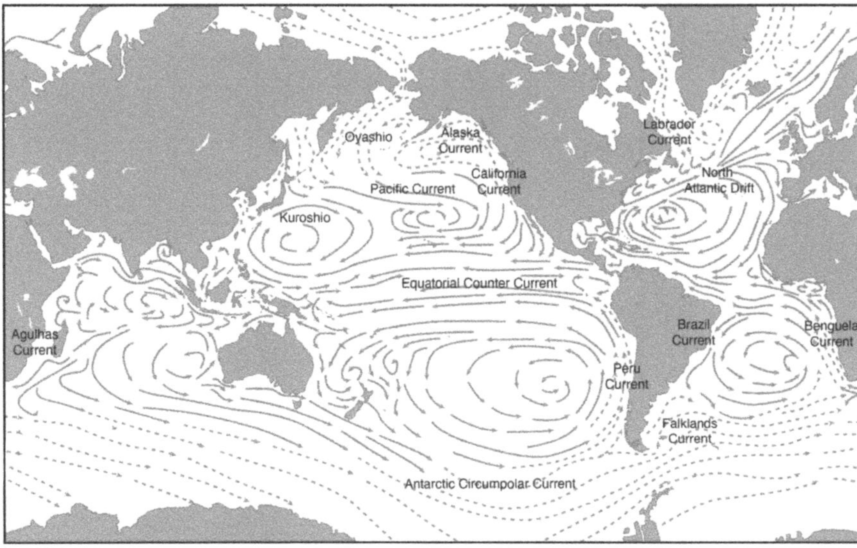

Figure 9. Major Surface Ocean Currents

hemispheres[4]. In fact, the deep ocean is the only candidate for driving and sustaining internal long-term climate change (of hundreds to thousands of years) because of its volume, heat capacity, and inertia. Today, the tropical Sun heats the surface water in the Gulf of Mexico. This heat also causes there to be a lot of evaporation sending moisture into the atmosphere starting the hydrological cycle. All this evaporation leaves the surface water enriched in salt. So, this hot salty surface water is pushed by the winds out of the Caribbean along the coast of Florida and into the North Atlantic Ocean. This is the start of the famous Gulf Stream (Figure 10). The Gulf Stream is about 500 times the size of the Amazon River at its widest point and flows along the coast of the USA and then across the North Atlantic Ocean, past the coast of Ireland, past Iceland, and up into the Nordic Seas. As the Gulf Stream flows northward, it becomes the North Atlantic Drift and it cools down. The combination of a high salt content and low temperature increases the surface water density or heaviness.

Let us now examine the difference between freshwater and seawater. As freshwater is cooled down, something amazing happens — it becomes denser down to a temperature of 4°C, after which it becomes lighter, and

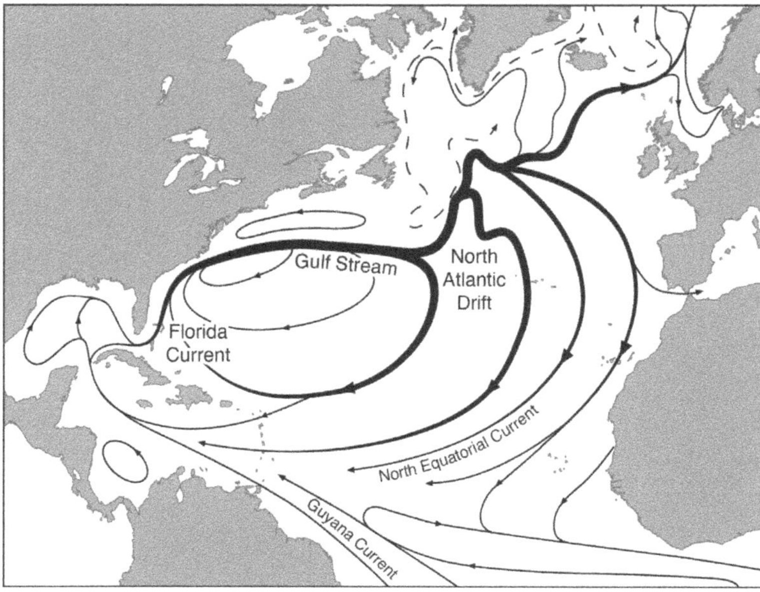

Figure 10. Major North Atlantic Surface Ocean Currents

then freezes at 0°C. This means that when ponds freeze, they do so from the top as the heaviest water sits on the bottom and is at 4°C, perfect for protecting any life within the pond or lake. As you progressively add salt to water, its freezing point drops, which is why we put salt on roads to stop them freezing, but also the temperature of greatest density drops (Figure 11). At 26 grams of salt per kilogram of water the temperature of greatest density and the freezing point coincide. This means seawater, which has 35 grams of salt per kilogram, will continue to get heavier and heavier until it freezes. When water freezes, another amazing thing happens — ice is formed, a solid that is lighter than its liquid form.

When the surface water reaches the relatively fresh oceans north of Iceland, the surface water has cooled sufficiently to become dense enough to sink into the deep ocean[7]. The 'pull' exerted by the sinking of this dense water mass helps maintain the strength of the warm Gulf Stream, ensuring a current of warm tropical water flowing into the northeast Atlantic, sending mild air masses across to the European continent. It has been calculated that the Gulf Stream delivers the same amount of energy as a million

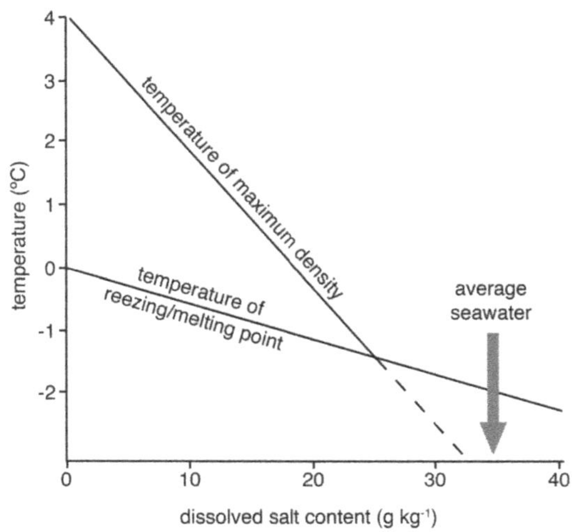

Figure 11. Temperature, Salt, and Density Relationship for Water

nuclear power stations. If you are in any doubt about how good the Gulf Stream is for the European climate, compare the winters at the same latitude on either side of the Atlantic Ocean, for example, London with Labrador, or Lisbon with New York. Or, better still, compare Western Europe and the west coast of North America, which have a similar geographical relationship between the ocean and continent — for example, Alaska and Scotland, which are at about the same latitude.

The newly formed deep water in the Nordic Seas sinks to a depth of between 2,000 metres and 3,500 metres in the ocean and flows southward down the Atlantic Ocean, as the North Atlantic Deep Water (NADW). In the South Atlantic Ocean, it meets a second type of deep water, which is formed in the Southern Ocean and is called the Antarctic Bottom Water (AABW). This is formed in a different way to NADW. Antarctica is surrounded by sea ice and deep-water forms in coast polynyas (large holes in the sea ice). Out-blowing Antarctic winds push sea ice away from the continental edge to produce these holes. The winds are so cold that they super-cool the exposed surface waters. This leads to more sea-ice formation and salt rejection because when ice is formed, it rejects any salt that the freezing water contains, which produces the coldest and saltiest water in the world. AABW flows around the Antarctic and penetrates the North

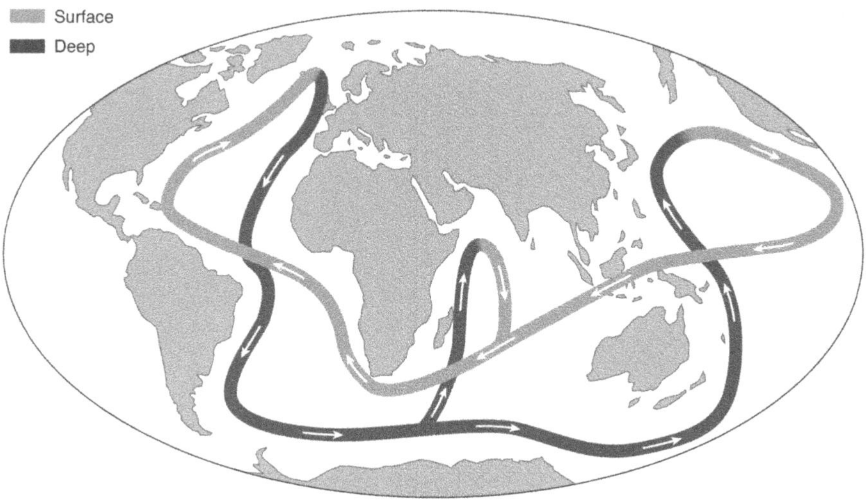

Figure 12. Global Deep-Ocean Conveyor Belt Circulation

Atlantic, flowing under the warmer and somewhat lighter NADW. The AABW also flows into both the Indian and Pacific Oceans. The NADW and AABW make up the key elements of the great global ocean conveyor belt (Figure 12), which allows heat to be exchanged between the two hemispheres on the timescale of hundreds and thousands of years.

The balance between the NADW and AABW is extremely important in maintaining our present climate, as not only does it keep the Gulf Stream flowing past Europe, it also maintains the right amount of heat exchange between the Northern and Southern Hemispheres. Scientists are worried that the circulation of deep water could be weakened or 'switched off' if there is sufficient input of fresh water to make the surface water too light to sink. This evidence has come from both computer models and the study of past climates and is discussed in detail in Chapter 8. Scientists have coined the phrase 'dedensification' to mean the removal of density by adding fresh water and/or warming up the water, both of which prevent seawater from being dense enough to sink. There is concern that climate change could cause parts of Greenland to melt. This could lead to more fresh water being added to the Nordic seas, thereby weakening the NADW and the Gulf Stream. This would bring much colder European winters with generally more severe weather. Since the influence of the warm Gulf

Stream is mainly in the winter, this change would not affect summer temperatures. So, if the Gulf Stream fails, global warming would still cause European summers to heat up. Europe would end up with extreme seasonal weather very similar to that of Alaska.

Summary

The climate system is very straightforward. It is controlled by more solar energy being received at the Equator than the Poles. Climate is simply the redistribution of energy due to this imbalance. It is the atmosphere and the oceans which undertake this redistribution. The ocean dominates heat exchange in the tropics while the atmosphere is more important in the mid to high latitudes. Complications are added because the Earth rotates every 24 hours plunging the Earth in and out of darkness. It also means the redistribution of energy away from the Equator takes place on a spinning ball. This creates the Coriolis effect and helps to explain why nearly all weather systems spin. Add to this the fact the Earth axis of rotation is at an angle with respect to the Sun, leading to a strong seasonal cycle. Heat is also redistributed between the hemispheres through the circulation of the deep ocean.

Chapter 2

What is Climate Change?

Introduction

Anthropogenic climate change is one of the defining challenges of the 21st century, along with poverty alleviation, environmental degradation, and global security. The problem is that 'climate change' is no longer just a scientific concern; it is now also of concern in terms of economics, sociology, geopolitics, national and local politics, law, and health, just to name a few. This chapter will examine briefly what 'anthropogenic' climate change is and the evidence that the global climate system is starting to change. The chapter will explain why changes in the climate system are causing unpredictable weather patterns, and an increase in the occurrence of extreme weather events, such as storms, floods, heatwaves, and droughts.

The Earth's Natural Greenhouse

The temperature of the Earth is determined by the balance between energy received from the Sun and its loss back into space. The Sun's energy consists of short-wave radiation (mainly visible 'light' and ultraviolet (UV) radiation) and nearly all of it passes through the atmosphere without interference (see Figure 13)[1]. The only exception is damaging high-energy UV, which is absorbed by atmospheric ozone. About one-third of the solar energy is reflected straight back into space while the remaining energy is absorbed by the surface of the Earth. This energy warms the land and the

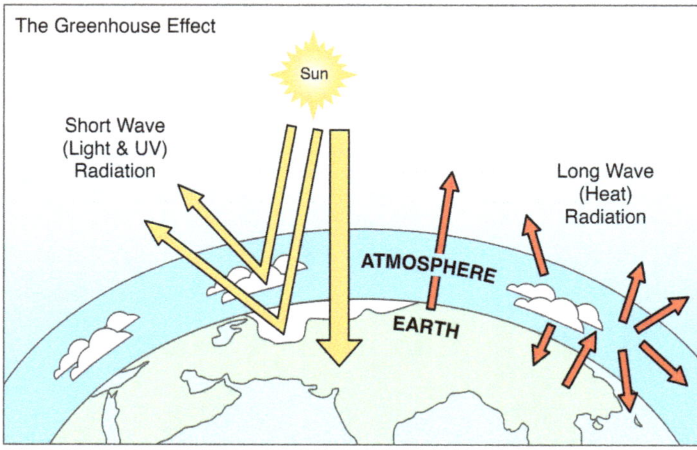

Figure 13. The Greenhouse Effect[1]

oceans, and this heat is radiated back as long-wave infrared or 'heat' radiation. Atmospheric gases such as water vapour (H_2O), carbon dioxide (CO_2), methane (CH_4), and nitrous oxide (N_2O) are known as greenhouse gases (GHGs) as they absorb some of this long-wave radiation, warming the atmosphere. This effect has been measured in the atmosphere and can be reproduced time and time again in the laboratory. Without this natural greenhouse effect, the Earth would be at least 35°C colder, making the average temperature in the tropics about −10°C. Since the Industrial Revolution, we have been burning fossil fuels (oil, coal, and natural gas) deposited hundreds of millions of years ago, releasing the carbon back into the atmosphere as CO_2 and CH_4, increasing the 'greenhouse effect', and elevating the temperature of the Earth. In effect, we are burning fossilised sunlight.

Climate Change is an Old Science

The history of climate change science is a long one and can be said to have started in 1856 when Eunice Newton Foote (American scientist, inventor, and women's rights campaigner) published her paper demonstrating the greenhouse effect of CO_2[2]. She used glass cylinders and mercury thermometers to show that when they were filled with different gases and placed in direct sunlight, the one containing carbon dioxide trapped the most heat. Looking to the history of the Earth, Foote theorised; "An atmosphere of that gas would give to our Earth a higher temperature."

Just three years later, John Tyndall, Professor of Natural Philosophy at the Royal Institution in London, demonstrated and measured the greenhouse effects of different gases[2]. Using apparatus that utilised thermopile technology, he was the first to correctly measure the relative infrared (heat) absorption of gases such as nitrogen (N), oxygen (O_2), water vapour (H_2O), carbon dioxide (CO_2), methane (CH_4) and ozone (O_3). He concluded that water vapour is the strongest absorber of radiant heat in the atmosphere and is the principal gas controlling air temperature of the Earth.

Building on the prior work of scientists such as Tyndall, Joseph Fourier, and Claude Pouillet, in 1896, the Swedish physical chemist Svante Arrhenius calculated how much the Earth's temperature would change due to variations in greenhouses gases. He estimated that a halving of atmospheric CO_2 would drop the Earth's temperature by 4°C and may have been a key cause of the ice ages. A doubling of CO_2, in turn, would increase the global temperature by 4°C. He concluded that anthropogenic carbon dioxide emission resulting from the burning of fossil fuels would be large enough to cause global warming[3].

But it wasn't until 1938 that the engineer and inventor Guy Stewart Callendar compiled 147 temperature records from around the world covering the previous 50 years and showed that the world was indeed warming (Figure 14). By using the few atmospheric carbon dioxide measurements available, he was able to suggest that a doubling of atmospheric carbon dioxide would cause a 2°C warming, half what had been suggested by Svante Arrhenius. Callendar's results were initially met with scepticism, but his papers, published through the 1940s and 1950s, prompted other scientists to investigate changes in atmospheric carbon dioxide and what was controlling them.

The Second World War saw a massive improvement in technology, including the development of infrared spectroscopy, and soon after the war, scientists showed that CO_2 in the upper atmosphere at low pressure did absorb heat — demonstrating the greenhouse effect. The worry about possible global warming was ignored at first as scientists argued that the oceans would simply absorb any extra anthropogenic carbon dioxide

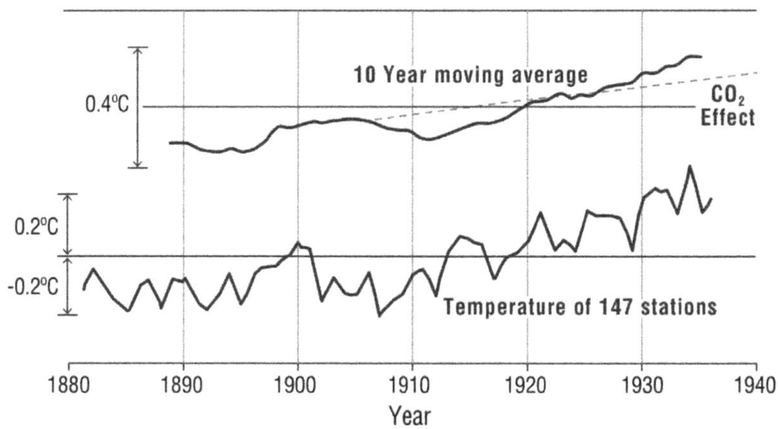

Figure 14. Guy Callendar's 1938 Global Temperature Compilation

emitted. Roger Revelle, director of the Scripps Institute of Oceanography in California, was concerned by this dismissal. Through his studies of surface ocean chemistry, he found that oceans return much of the CO_2 that they absorb back into the atmosphere[3]. This was a great revelation, and showed that because of the peculiarities of ocean chemistry, the oceans would not be the complete sink for anthropogenic CO_2, as was first thought. We now know that the oceans are taking up about one-quarter of the annual total anthropogenic production (Figure 15).

Charles Keeling, who was hired by Roger Revelle, made the next important step forward in climate change science. In the late 1950s and early 1960s, Keeling used the most modern technology available to measure the concentration of atmospheric CO_2 in Antarctica and Mauna Loa. He showed clear evidence that levels of atmospheric CO_2 have been rising ever since the beginning of the Industrial Revolution. The first measurements of CO_2 concentrations in the atmosphere started in 1958, on the summit of Mauna Loa Mountain in Hawaii at an altitude of about 4,000 m. The measurements were made at this remote location to avoid contamination from local pollution sources. The record clearly shows that atmospheric concentrations of CO_2 have increased every single year since the first measurement in 1958[3]. The mean concentration of approximately 316 parts per million by volume (ppmv) in 1958 has risen to over 420 ppmv today (see Figure 16). The annual variations in the Mauna Loa

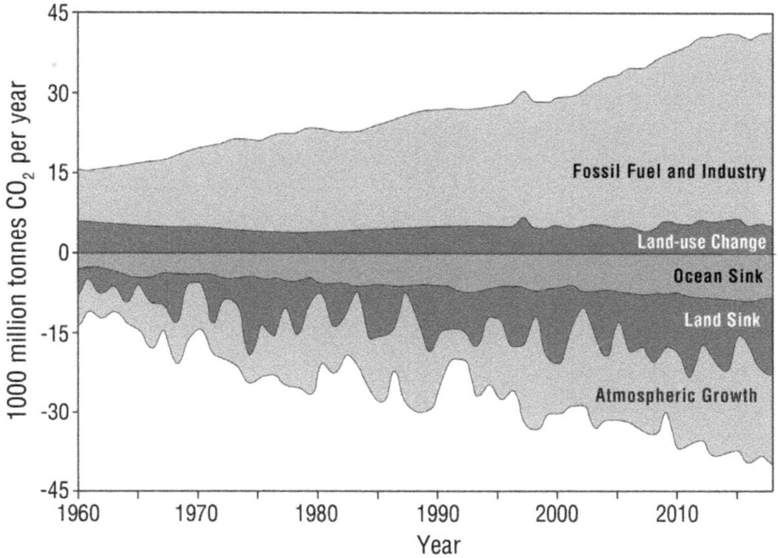

Figure 15. Historic Global Sinks and Sources of Carbon Dioxide

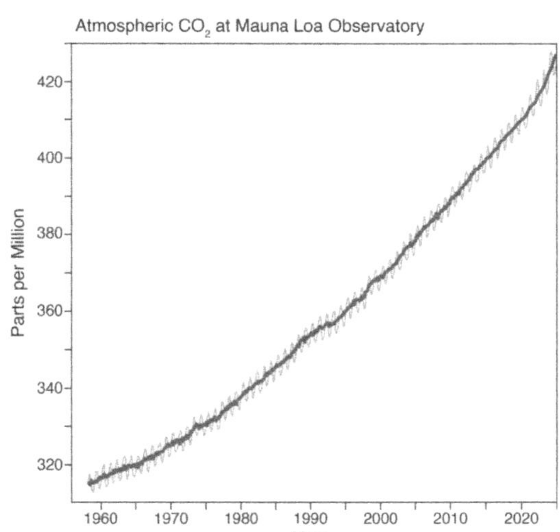

Figure 16. Keeling Curve: Mauna Loa (Hawaii) Observatory Record of Atmospheric Carbon Dioxide.

observatory are mostly due to CO_2 uptake by growing plants. The uptake is highest in the Northern Hemisphere springtime due to the great expanse of land and hence every spring, there is a drop in atmospheric CO_2, which unfortunately does nothing to change the overall trend towards ever higher values.

Why the Delay in Recognising Climate Change?

In 1959, the physicist Gilbert Plass published an article in *Scientific American* declaring that the world's temperature would rise by 3°C by the end of the century. The magazine editors published an accompanying photograph of coal smoke belching from factories and the caption read, 'Man upsets the balance of natural processes by adding billions of tons of carbon dioxide to the atmosphere each year.' This resembles thousands of magazine articles, television news items, and documentaries that we have all seen since the late 1980s. So, why was there a delay between the science of global warming being accepted in the late 1950s and the realisation by those outside the scientific community of the true threat of global warming at the beginning of the 21st century?[3]

The key reasons for the delay in recognising climate change were the lack of increase in global temperatures and the lack of global environmental awareness. The global mean temperature (GMT) data set is calculated by compiling all the available land and sea temperatures. From 1940 until the mid-1970s, the global temperature curve seems to have had a slight downward trend (Figure 17). This provoked many scientists to discuss whether the Earth was entering the next great ice age. Increasing knowledge of past climates in the 1970s and 1980s showed that this was highly unlikely, as ice ages take thousands of years to form. Even so, it was not until the late 1980s, when the annual GMT curve started to increase, that the global cooling scenario was finally discarded. By the late 1980s, the annual GMT curve was rising so steeply that all the dormant evidence from the late 1950s and 1960s was dusted off and the global warming theory came to prominence. In fact, it was in 1988 that Professor Jim Hansen, the director of the National Aeronautics and Space Administration (NASA) Goddard Institute for Space Studies, was asked to testify on the

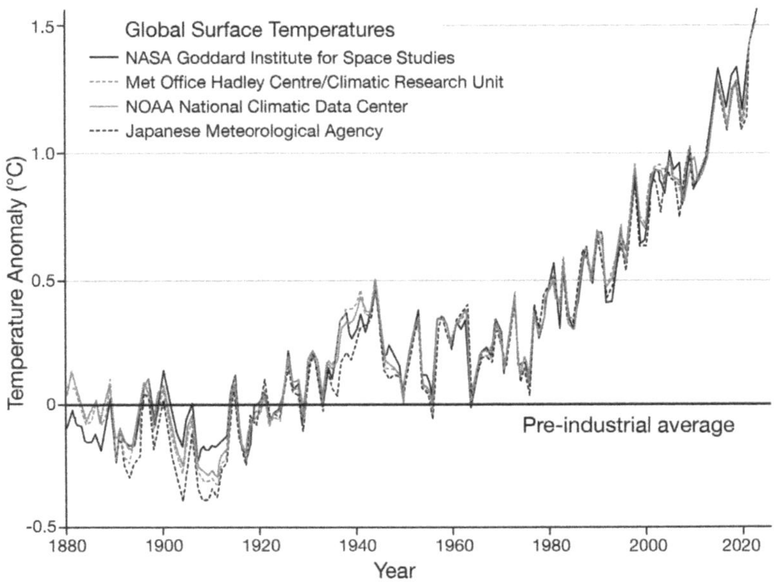

Figure 17. Variation of the Earth's Surface Temperature Over the Last 150 Years

matter before the United States Senate Committee on Energy and Natural Resources. He stated that, "Global warming has reached a level such that we can ascribe with a high degree of confidence a cause-and-effect relationship between the greenhouse effect and observed warming...It is already happening now." This testimony was widely reported by the media, and global warming became a mainstream issue[3].

It seems then that the eventual recognition of climate change was driven by the upturn in the annual GMT data set. The latest Intergovernmental Panel on Climate Change (IPCC) 2021 Science Report has reviewed and synthesised a wide range of data sets. It shows that the trend in global temperature first recognised in the late 1980s is correct, and that this warming trend has continued unstopped until the present day (see Figure 17)[1].

The upturn in the annual GMT data was not the sole reason for the new prominence of the global warming issue. In the late 1970s and 1980s, there were significant advances in global climate modelling. These new ocean–atmosphere-coupled general circulation models (GCM) produced estimates of significant warming associated with a doubling of CO_2 in the atmosphere — closer in fact to Arrhenius's original calculations. By the 1980s, scientific concern had emerged about CH_4 and other non-CO_2 GHGs, as well as the role of the oceans as a carrier of heat. GCMs continued to improve, and the numbers of scientific teams working on such models increased over the 1980s and the 1990s. In 1992, a first overall comparison of results from 14 GCMs was undertaken; the results were all in rough overall agreement, confirming that rising GHGs would cause significant global warming. In the latest IPCC sixth assessment report in 2021, over 100 distinct supercomputer climate models were used, produced by 49 different international modelling groups, all of which confirm GHGs are warming up our planet (see Chapter 6)[1].

Scientists using the latest climate models predict that if we continue on our current carbon emissions pathway, we could warm the planet by between 1.5 and 5.5°C compared with the pre-industrial period (1850–1900) by the end of the century. Economists warn that this could cost as much as 20% of the world's gross domestic product (GDP). In the face of such a threat, it is essential to understand the history of climate change and the evidence that supports it.

Summary

The Earth is at least 35°C warmer than it should be due to natural levels of greenhouse gases. The most important natural GHGs in order of importance are H_2O, CO_2, and CH_4. We have known about the greenhouse effect for over 170 years. Since the industrial revolution, we have been adding additional GHGs to the atmosphere. Global warming due to these extra gases was theorised in 1896 by Svante Arrhenius and shown to be occurring in 1938 by Guy Callendar. However, there was a delay in recognising the threat of climate change until the late 1980s when data clearly showed the world was warming. This warming has continued as GHG emissions have increased every single year, and global warming is now over 1.5°C above pre-industrial levels.

Chapter 3

Weather Versus Climate

Introduction

Many people get weather and climate confused. This confusion is exacerbated when scientists are asked to predict climate 50 years from now when everyone thinks we cannot predict the weather a few weeks ahead. Climate is generally defined as 'the average weather'. The original definition of climate was 'the average weather over 30 years', this has been changed because we now know that our climate is changing rapidly and significant changes have been seen every decade for the last 50 years. The chaotic nature of the weather can make it unpredictable beyond a few days, while understanding the climate and modelling climate change is much easier as you are dealing with long-term averages. A good comparison is that of a person's life expectancy — though it is impossible to predict at what age any particular person will die, we can say with a high degree of confidence that the average life expectancy of a person in a developed country is about 80 years. The other confusion is that people always remember extreme weather events but not the average weather. For example, everyone remembers the extreme heatwave in China in 2022 when over 50,000 people died or the big freeze of Beijing with nearly two weeks of temperatures below –10°C, or the devastating wildfires in Australia in 2020 that killed nearly three billion animals or the floods in Pakistan in 2022 which affected 20 million people. Our perception of weather is skewed by these events rather than by an appreciation of the average weather or climate.

Chaos Theory

The National Weather Service in the USA spends over $1 billion per year ensuring the country has the most accurate weather prediction possible. In other countries, similar resources are poured into weather agencies, as predicting the weather is big business and getting a storm prediction right can save billions of dollars and many lives. Today, four-day forecasts are as accurate as the two-day forecasts were 20 years ago. Predictions of rain in three days' time are as accurate as one-day forecasts were in the mid-1980s. The accuracy of flash flood forecasts has improved from 60% to 86% correct; moreover, potential victims of these floods get nearly an hour's warning instead of the 8 minutes they would have had in 1986. Tornadoes warnings have improved from 5 minutes in 1986 to over 13 minutes now. Severe local thunderstorms and similar cloudbursts are typically seen 18 minutes beforehand rather than 12 minutes over two decades ago. 75% of all hurricane paths can be predicted at least 24 hours in advance and the landfall of a hurricane can be predicted to within 100 miles (160 km) which is impressive given they are between 300 to 400 miles wide.

These are great achievements, but it does not explain why with all our technology and our understanding of the climate system, we cannot predict the weather 10 days, a month, or a year in advance. Think of all those times that the weather report on television has said it will be sunny today and then it rains. So why is it so difficult to predict the weather? In the 1950s and 1960s, it was thought that our weather prediction was limited by our lack of data and that if we could measure things more accurately and clearly understand the fundamental processes, we would be able to achieve a much higher level of prediction. But in 1961, Edward Lorenz, a meteorologist at Massachusetts Institute of Technology, made a cup of coffee that radically changed the way we think about natural systems. In 1960, Lorenz had produced one of the first computer models of weather. One day in the winter of 1961, Lorenz's computer model produced some very interesting patterns, which he wanted to look at in greater detail[1]. As computers took a longtime to program, he took a short-cut and decided to start mid-way through the run. As this was one of the earliest computers, he had to retype all the starting numbers. Instead of typing them into six decimal places (e.g., 0.506127), he only typed the first three to save time and space, and then went and made that famous cup of coffee. When Lorenz came back, he found that the weather patterns had diverged from the initial run so much that there was no recognisable similarity between them. It seems the model was very sensitive to the very small changes, that one part in a thousand instead of being inconsequential had had a huge effect on the outcome. This original work has led to the development of chaos theory. Chaos theory shows us that very small variations in atmosphere temperature, pressure, and humidity can have a major and unpredictable or chaotic effect on large-scale weather patterns[2].

Nevertheless, chaos theory does not mean there is a complete lack of order within a system[1]. Far from it, chaos theory tells us that we can predict within certain boundaries what the weather will be like: we all know, for example, that most tornadoes occur in May in the USA and that winters are wet in England. But when it comes to more detailed prediction, everything breaks down due to what has become known as the 'butterfly effect'. The idea is that small changes represented by the flapping of the wings of a butterfly can have a large effect on the weather, for example: altering the strength and direction of a hurricane. Errors and uncertainties multiply and cascade upward through the chain of turbulent features from dust devils and squalls up to continental size eddies that only satellites can see. In effect, we will never know which of the small weather changes will combine to have these large effects. While Lorenz used 12 equations in his weather model, modern weather computers use over 500,000. But even the best forecasts, which come from the European Centre for Medium Weather Forecasts based at Reading in England, suggest that weather predictions for more than four days are at best speculative and beyond a week worthless, all because of chaos.

Chaos theory says that we can understand weather and we can predict general changes but it is very difficult to predict individual events such as rainstorms and heatwaves[1]. The study of climate, however, has one great advantage over meteorology because it only examines averages and thus, chaos theory does not affect it. Moreover, when it comes to modelling future climate change, we can now understand that an increase in the Earth's average temperature will make some weather phenomena more frequent and intense, for example: heatwaves and heavy rainfall events, while others will become less frequent and intense, for example: extreme cold events and snow fall (see Chapter 6).

Decadal and Quasi-Periodic Climate Systems

The climate system contains many cycles and oscillations that complicate our ability to predict the weather. These include decadal cycles such as the North Atlantic oscillation (NAO), Arctic oscillation (AO), and the Pacific Decadal Oscillation (PDO). The first of these is the NAO[3], which was first described in the 1920s by Sir Gilbert Walker (14 June 1868–4 November 1958), a British physicist and statistician. The NAO is a climate phenomenon in the North Atlantic Ocean and is represented by the atmospheric pressure difference at sea level between Iceland and the Azores. The difference in Icelandic low-pressure and the Azores high-pressure systems seems to control the strength and direction of westerly winds and storm tracks across the North Atlantic Ocean. This in turn controls where and when in Europe it rains. Unlike the El Niño–Southern Oscillation, the NAO is largely controlled by changes in the atmosphere. The NAO is closely related to the AO and though both seem to change on a decadal scale, there seems to be no periodicity. The PDO[4] is a long-term pattern of climate variability in the Pacific Ocean, characterised by varying sea surface temperatures over periods of 20–30 years. It has two main phases; a positive (or warm) phase and a negative (or cool) phase affecting weather patterns across the Pacific region. It is also known to interact with El Niño–Southern Oscillation (ENSO).

El Niño–Southern Oscillation

One of the most important and mysterious elements in global climate is the periodic switching of direction and intensity of ocean currents and winds in the Pacific. Originally known as El Niño ('Christ child' in Spanish) as it usually appears at Christmas, and now more often referred to as ENSO[5], this phenomenon typically occurs every 3 to 7 years. It may last from several months to more than a year. ENSO is an oscillation between three climates: the 'normal' conditions, La Niña, and 'El Niño'.

ENSO has been linked to changes in the monsoon, storm patterns, and occurrence of droughts throughout the world. For example, a prolonged ENSO event can cause severe climate changes all over the Earth including droughts in East Africa, northern India, north-east Brazil, Australia, Indonesia, and Southern USA; and heavy rains in California, parts of South America, the Pacific, Sri Lanka, and east central Africa. The state of the ENSO has also been linked into the position and occurrence of hurricanes in the Atlantic Ocean.

An El Niño event is when the warm surface water in the western Pacific moves eastward across to the centre of the Pacific Ocean (Figure 18). Hence, the strong convection cell or warm column of rising air is much closer to South America. Consequently, the Trade Winds are much weaker and the ocean currents crossing the Pacific Ocean are

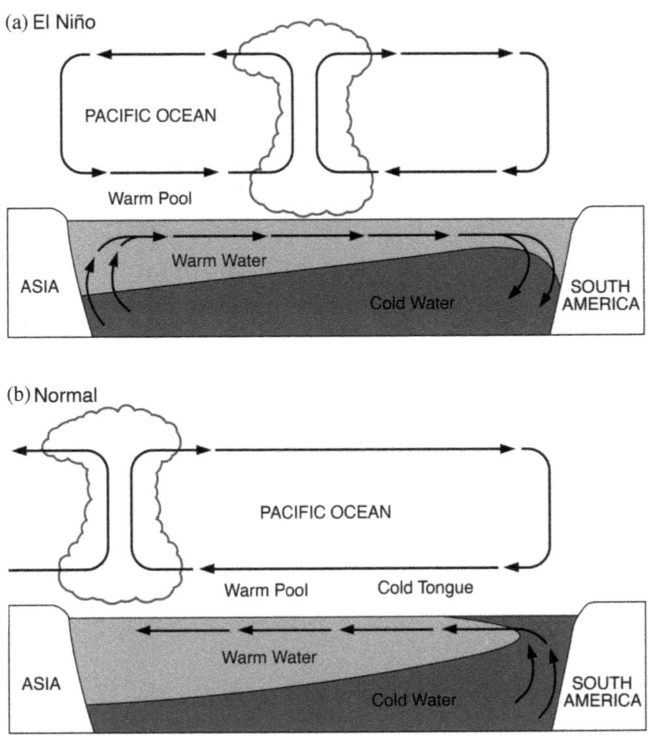

Figure 18. Pacific Ocean during El Niño and Normal Periods

weakened. This reduces the amount of the cold, nutrient-rich upwelling off the coast of South America and without those nutrients, the amount of life in the ocean is reduced and fish catches are dramatically reduced. This massive shift in ocean currents and the position of the rising warm air changes the direction of the jet streams that upset the weather in North America, Africa, and the rest of the world. However, if you ask what causes El Niño, then the answer is of the chicken and egg variety. Does the westward ocean current across the Pacific reduce in strength, allowing the warm pool to spread eastward and moving with it the wind system? Or does the wind system relax in strength, reducing the ocean currents, and allowing the warm pool to move eastwards? Many scientists believe that long period waves in the Pacific Ocean that move between South America and Australia over time help shift the ocean currents, which produce either an El Niño or a La Niña period.

La Niña is a more extreme version of the 'normal' conditions[5]. Under normal conditions, the Pacific warm pool is in the western Pacific and there are strong westerly winds and ocean currents keeping it there. This results in upwelling off the coast of South America, providing lots of nutrients and thus creating excellent conditions for fishing. During a La Niña period, the temperature difference between the western and eastern Pacific becomes extreme and the westerly winds and ocean currents are enhanced. La Niña impacts on the world's weather are less predictable than those of El Niño. This is because during an El Niño period, the Pacific jet stream and storm tracks become strong and straighter, making it easier to predict its effects. La Niña, on the other hand, weakens the jet stream and storm tracks, making them loopier and irregular, meaning that the behaviour of the atmosphere and particularly of storms becomes more difficult to predict. In general, where El Niño is warm, La Niña is cool, where El Niño is wet, La Niña is dry.

Monsoons

The other important area for massive rainstorms is the monsoon belt. The name monsoon comes from the Arabic word 'mausim' which means 'season', as most of the rains that fall in Southeast Asia occur during the summer. In the tropics, the sun's energy is most intense as the sun is directly overhead. This heats up the land and sea, thus warming the air above. This warm, moist air rises, leaving an area of low pressure beneath it, which helps to suck in air from the surrounding area (Figure 19). This suction results in the Trade Winds, which can travel from much higher latitudes to this area of rising air. As the winds come from both the Northern and Southern Hemispheres, this area is known as the Intertropical Convergence Zone (ITCZ). As the air at the ITCZ rises, it forms huge, towering clouds and produces large amounts of rain. The ITCZ moves north and south with the seasons as the position of the most intense sunlight shifts up and down across the Equator. It is also influenced strongly by the position of the continents. This is because the land heats up faster and to a greater extent than the ocean, thus pulling the ITCZ even further north or south during that season. An example of this is the Asian summer monsoon: during the summer, near the Himalayan mountains and the lowlands of

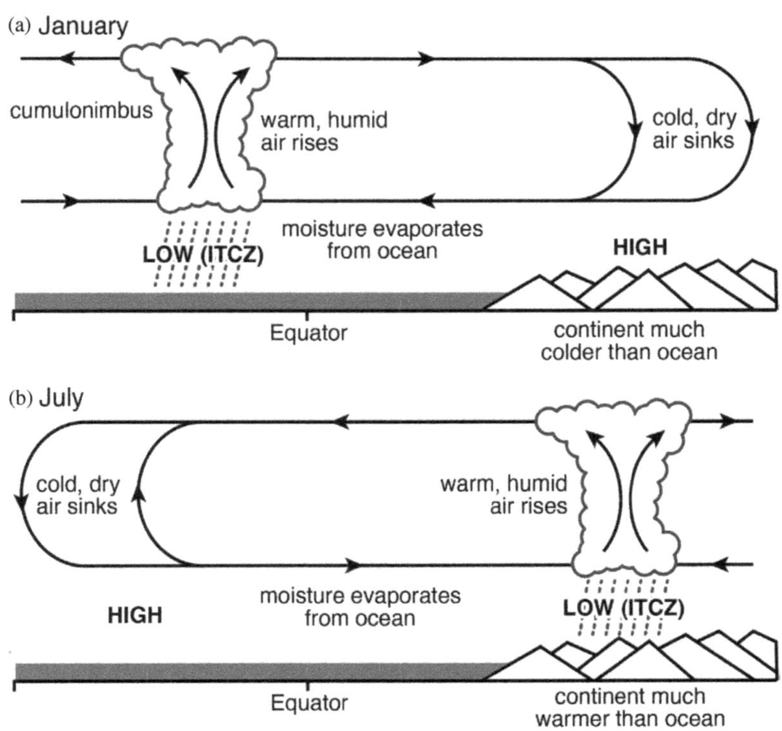

Figure 19. Monsoonal System

India heat up. This pulls the ITCZ across the Equator on to Asia. Because the Southern Hemisphere winds have been pulled across the warm Indian Ocean, they are warm and full of moisture; when they are forced to rise and cool down over India, they produce very heavy rainfall throughout Southeast Asia and as far north as Japan. During Northern Hemisphere winter, the ITCZ moves south of the Equator, but in Southeast Asia, it means warm, moist winds from the North Pacific are dragged southward across the continent into the Southern Hemisphere. This means that some areas such as Indonesia and Southern China get two monsoonal rainy seasons a year: one from the north and one from the south. No wonder this is the most fertile place on Earth, supporting over two-fifths of the world's population. Despite being the bringers of life, the rains can cause catastrophic hazards, especially in the form of flooding. Examples of this are the terrible floods in 1998 in Bangladesh and China, which caused over $30 billions of damages and thousands of deaths[6].

Amazon Monsoon

During the Southern Hemisphere summer, the continent of South America heats up. This rising air leaves an area of low pressure at ground level, which is filled by sucking in the surrounding air. This pulls the ITCZ between the North and Southern tropical air southward over Brazil. The southward shift of the ITCZ brings with it lots of rain as the air being pulled across the Equator from the north originates over the warm, tropical Atlantic Ocean. This produces the Amazon monsoon and results in the mightiest river in the world, as well as the greatest extent of rainforest on the planet. The Amazon Basin covers an amazing 2.7 million miles^2, much of which is covered with rainforest. The Amazon River delivers 20% of all the freshwater that enters the world's oceans. Without the monsoon rains, the most diverse habitat in the world would not exist.

Living Under the Asian Monsoons

The fertility of Southeast Asia and the reason why it supports half the population of the world is in part due to the life bringing Monsoons. But they also bring damage and risks. A good illustration of this is

Bangladesh — a country literally built by the monsoons as over three-quarters of the country is a deltaic region formed by the sediments brought in by the Ganges, Brahmaputra, and Meghna rivers — all fed by the summer monsoons. Over half the country lies less than 5 metres above sea level, thus flooding is a common occurrence. During a normal summer monsoon, a quarter of the country is flooded. Yet these floods, like those of the Nile, bring life with them as well as destruction. The water irrigates and the silt fertilises the land. The fertile Bengal Delta supports one of the world's most dense populations, over 110 million people in 140,000 km^2. But every so often, the monsoon floods exceed what even Bangladesh can cope with. In 1998, three-quarters of the country was flooded for 2 months, causing billions of pounds worth of damage and thousands of deaths. Bangladesh also copes with tropical cyclones but the good news is that the loss of life has dramatically dropped. In 1970, there were over 300,000 cyclone-related deaths; in 1991, there were 138,000; in 2007, Cyclone Sidr killed 4,234; in 2009, Cyclone Aila killed 190 and in 2024, Cyclone Remal resulted in around 65 deaths. This is not because the tropical cyclones have grown gentler, far from it. It is because of good governance. The Bangladesh government has, first, invested in excellent meteorological facilities to make as accurate a prediction of when and where the cyclones will make land fall; second, they have set up a communication network using cyclists, so that once a cyclone warning is given, the message is carried to all the towns and villages that will be affected. They have also built cyclone shelters, protected water and sanitation facilities, and encouraged floating agriculture, which can withstand the storms. These relatively simple changes have resulted in the saving of hundreds of thousands of lives[7].

Summary

Weather is the day-to-day or even hour-to-hour state of the atmosphere, whereas climate is the consistent patterns of weather over a much longer period. For example, how warm, dry and cloudy it is on a summer's day is weather, but the fact that it is on average, warm, dry and sunny in summer is climate. One way to think of it is weather is what you wear today, and climate is the wardrobe that you own.

Chapter 4

Past Climate Changes

Introduction

Understanding climate change in the past is essential as it provides deep insights to how the climate changes and what the major mechanisms are. It also provides insights to how our climate will change in the future due to human impacts. Climate change in the geological past is reconstructed using a number of key archives, including marine and lake sediments, ice cores, cave deposits, and tree rings. These various records show that Earth's climate over the last few billion years has varied frequently from a deep freeze to a tropical paradise[1]. But more relevant to future climate is the fact that over the past 50 million years, the Earth's climate has been cooling down, moving from the so-called 'greenhouse world' of the Eocene, with warm and gentle conditions, through to the cooler and more dynamic 'icehouse world' of today. It may seem odd that in geological terms, our planet is currently extremely cold, while we worry about global warming. This is because the very fact that there are huge ice sheets on both Antarctica and Greenland, and nearly permanent sea ice in the Arctic Ocean, makes the global climate very sensitive to changes in GHGs.

Icehouse and Greenhouse Worlds

Plate tectonics drives the slow shift of the continents across the globe, shifting from a supercontinent to fragmented continents and then back again. The supercontinent Rodinia formed about 1.1 billion years ago and broke up roughly 750 million years ago. One of the fragments included large parts of the continents we now find in the Southern Hemisphere. Plate tectonics brought the fragments of Rodinia back together in a different configuration about 300 million years ago, forming the best-known supercontinent, Pangaea. Pangaea subsequently broke up into the northern and southern supercontinents of Laurasia and Gondwana, about 200 million years ago. Both of these supercontinents have continued to fragment over the last 100 million years. Icehouse climates form when the continents are moving together (Figure 20)[2]. The sea level is low due to lack of seafloor production. The climate becomes cooler and more arid, because of the reduction in rainfall due to the strong rain shadow effect of large super-plateaus. Greenhouse climates, on the other hand, are formed as the continents disperse, with sea levels high due to the high level of sea floor spreading. There are relatively high levels of carbon dioxide in the atmosphere, possibly over three times the current levels, due to production at oceanic rifting zones. This produces a warm and humid climate.

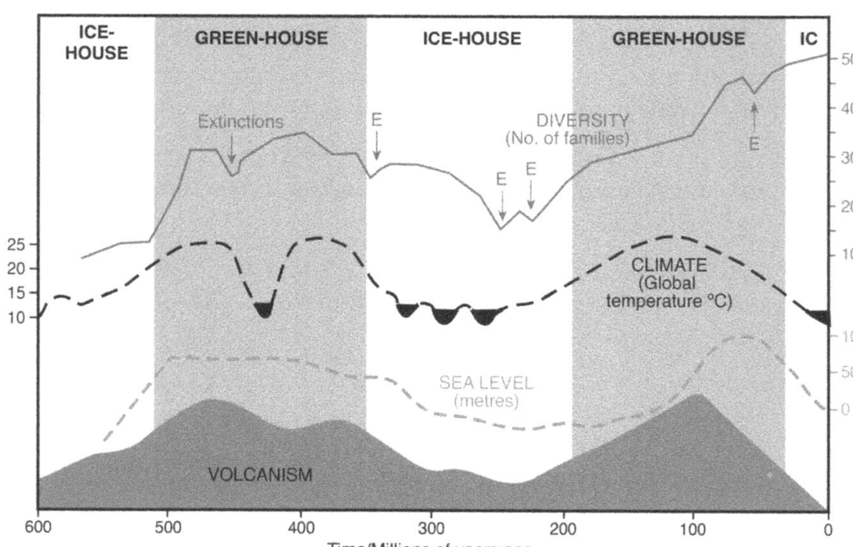

Figure 20. Relationship between Icehouse and Greenhouse World and Tectonics, Sea Level, Climate and Extinctions

The formation and break up of these supercontinents had a huge effect on evolution. Supercontinents are extremely bad for life. First, there is a massive reduction in the amount of shelf sea areas, where we think multi-cellular life may have started. Second, the interiors of continents are very dry and global climate is usually cold. A number of key mass extinctions are correlated with the formation of supercontinents. For example, it is estimated that up to 96% of all marine species and 70% of terrestrial vertebrate species became extinct during the Permian–Triassic extinction event 250 million years ago, which is nicknamed the 'mother of all mass extinctions. It is also not surprising that the explosion of complex, multi-cellular organisms occurred during the Cambrian period about 550 million years ago, following the break-up of the Rodinia supercontinent.

Snowball Earth

There is a theory that prior to about 650 million years ago, the surface of the Earth became entirely frozen at least once — the so-called Snowball Earth hypothesis[1]. It is a way to explain the sedimentary deposits found in the tropics, which show glacial features that suggest there must have been a lot of ice in the tropics. Opponents of the idea suggest that the geological evidence does not suggest a global freezing. Moreover, there is difficulty in getting the whole ocean to become ice- or even slush-covered. There is also the difficulty of seeing how the world, once in a snowball state, would subsequently escape the frozen condition. One answer is that this would occur through the slow build-up of atmospheric carbon dioxide and methane, which would eventually reach a critical concentration, warming the atmosphere enough to start the melting process. There are a number of unanswered questions, including whether the Earth was a *full* snowball or a 'slush ball' with a thin equatorial band of open water. But what is particularly interesting is the idea that the evolution of complex life put an end to the possibilities of ever having a snowball Earth again. Professor Andy Ridgwell at Bristol University has suggested that the evolution of marine micro-organisms that form calcite shells now buffers the oceans' carbonate system so much that the extreme variation in atmospheric carbon dioxide needed to plunge the world into or out of a snowball or slush ball condition could not now occur.

Our Current Icehouse World

The long-term global cooling of the Earth kicked off with the glaciation of Antarctica about 35 million years ago and then accelerated with the great Northern Hemisphere ice ages, which began 2.5 million years ago[3]. Since the beginning of the great ice ages, the global climate has cycled between conditions that were similar or even slightly warmer than today, to full glacial phases, during which ice sheets over 3 km thick formed over much of North America and Europe (Figure 21). Between 2.5 and 1 million years ago, these glacial–interglacial cycles occurred every 41,000 years, and since 1 million years ago, they have occurred about every 100,000 years. These great ice-age cycles are driven primarily by changes in the Earth's orbit with respect to the Sun. In fact, the world has spent over 80% of the past 2.5 million years in conditions colder than the present. Our present interglacial, the Holocene Period, began about 10,000 years ago, and is an example of the brief warm conditions that occur between each ice age. The Holocene began with the rapid and dramatic end of the last ice age: in less than 4,000 years, global temperatures increased by 6°C, global sea level rose by 120 metres, atmospheric CO_2 increased by one-third, and atmospheric CH_4 doubled. Still, this is much slower than the changes we are seeing today[3].

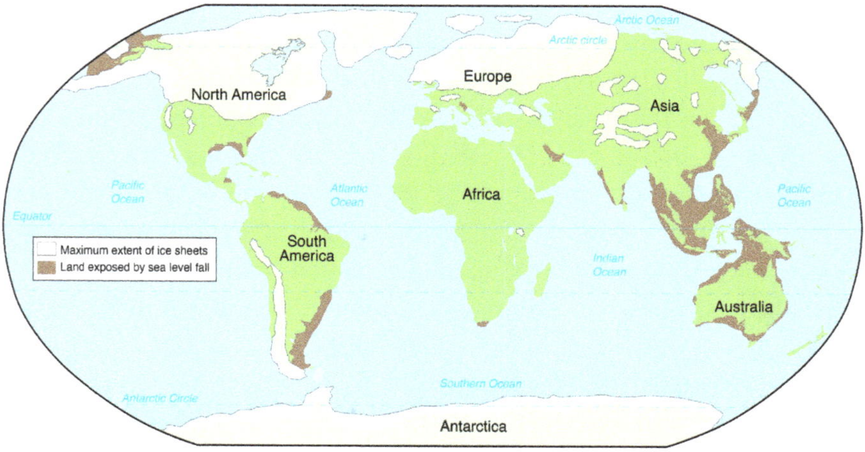

Figure 21. Global Map of Last Ice Age

Waxing and Waning of the Great Ice Ages

We now know that glacial–interglacial cycles are the fundamental characteristic of the Quaternary Period, the last 2.5 million years. The waxing and waning of the huge continental ice sheets is initiated by the changes in the Earth's orbit around the sun. The Earth over long periods of time wobbles on its axis; that changes the amount of sunlight or solar energy received by different parts of the Earth. These small changes are enough to push or force climate change. These waxing and waning are not caused by Earth's orbital wobbles, but rather by the Earth's climate reaction, which translates relatively small changes in regional solar energy into major climatic variability. For example, the position of the Earth orbit is very similar today to what it was 21,000 years ago during the last ice age. It is not the exact orbital position that controls the climate but rather the changes in the orbital positions. There are three main orbital parameters or wobbles, called eccentricity, obliquity or tilt, and precession[4]. Each has a unique cycle and effect on climate and when combined, they can push the climate either into or out of a great ice age.

Orbital Forcing

Eccentricity is the shape of the Earth's orbit around the sun, which varies from a circle (precession) to an ellipse (Figure 22). These variations occur over a period of about 96,000 years with an additional long cycle of about

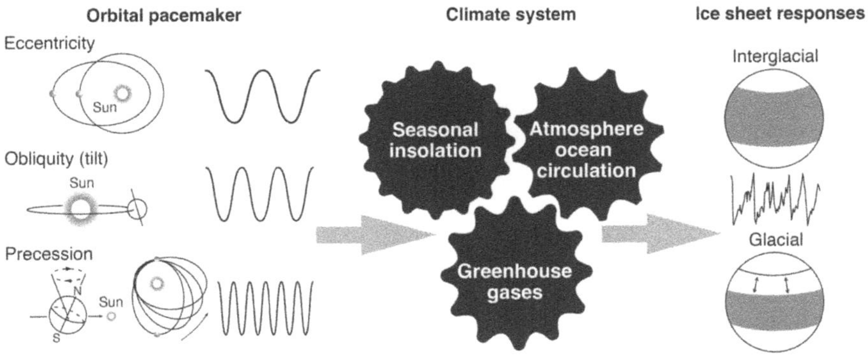

Figure 22. Orbital Variables

400,000 years. Described another way, the long axis of the ellipse varies in length over time. In recent times, the Earth is closest to the sun on 3 January at about 146 million km; this position is known as perihelion. On 4 July, the Earth is at its greatest distance from the sun, at about 156 million km; this position is known as the aphelion. Changes in eccentricity cause only very minor variations in total insolation, but can have significant seasonal effects when combined with precession. If the orbit of the Earth were perfectly circular, there would be no seasonal variation in solar insolation. Today, the average amount of radiation received by the Earth at perihelion is ~351 Wm^2 and is 329 Wm^2 at aphelion. This represents a difference of ~6%, but at times of maximum eccentricity (ellipse length) over the last 5 million years, the difference could have been as great as 30%. Milutin Milankovitch in 1949 suggested that northern ice sheets are more likely to form when the sun is more distant in summer, so that each year, some of the previous winter's snow survives. The other effect of eccentricity is to modulate the effects of precession. It is important to note that eccentricity is by far the weakest of all three orbital parameters.

Obliquity or the tilt of the Earth's axis of rotation with respect to the plane of its orbit varies between 21.8° and 24.4° over a period of 41,000 years. It is the tilt of the axis of rotation that causes the seasons, as described in Chapter 1. The larger the obliquity, the greater the difference between the insolation received in summer and winter. Milutin Milankovitch suggested that the colder the Northern Hemisphere summers, the more likely it is that snowfall will accumulate, resulting in the gradual build-up of glaciers and ice sheets.

Precession has components that relate to the elliptical orbit of the Earth (eccentricity) and its axis of rotation. The Earth's rotational axis precesses every 27,000 years. This is similar to the gyrations of the rotational axis of a toy spinning top. Precession causes a change in the Earth–sun distance for any particular date, for example, the beginning of the Northern Hemisphere summer. It is the combination of the different orbital parameters that lead to the two different precessional periodicities of 23,000 years and 19,000 years. Combining the precession of the axis of rotation plus the precessional changes in orbit produces a period of 23,000 years. However, the combination of eccentricity (96,000 years) and precession of the axis of rotation also results in a period of 19,000 years.

These two periodicities combine so that perihelion coincides with the summer season in each hemisphere on average every 21.7 thousand years. Precession has the most significant impact in the tropics (in contrast to the impact of obliquity at the Equator, which is zero). Although obliquity clearly influences high latitude climate change, which may ultimately influence the tropics, the direct effects of insolation in the tropics are due to eccentricity-modulated precession alone. Combining the effects of eccentricity, obliquity, and precession provides the means for calculating insolation for any latitude through time.

Clockwork Climate

Combining the effects of all three orbital parameters, you can calculate the solar energy received for any latitude back through time. In 1949, Milutin Milankovitch, a brilliant Serbian mathematician and climatologist suggested that summer insolation at 65°N which is just south of the Arctic Circle was critical in controlling glacial–interglacial cycles (Figure 23)[4]. He argued that if the summer insolation was reduced significantly to allow

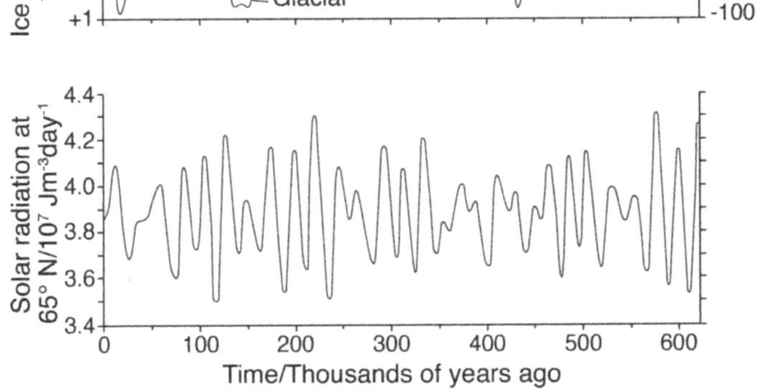

Figure 23. Comparison of Northern Hemisphere Solar Radiation and Global Sea Level

ice to survive through the summer, it would start to build up, and eventually produce an ice sheet. Orbital forcing does have a large influence on this summer insolation; the maximum change in solar radiation in the last half million years is equivalent to reducing the amount of summer radiation received today at 65°N to that received now over 550 km to the north at 77°N. In simplistic terms, this brings the current ice limit in mid-Norway down to the latitude of mid-Scotland. The reason why it is at 65°N and not at 65°S, which controls climate, is very simple. Any ice that builds up in the Northern Hemisphere has many continents to grow upon. In contrast, in the Southern Hemisphere, the ice growth is limited by the Southern Ocean, as any extra ice produced on Antarctica falls into the ocean and is swept away to warmer seas. The conventional view of glaciation is that low summer solar energy in the temperate Northern Hemisphere allows ice to survive summer, and thus ice sheets start to build up on the northern continents. But this apparently simple clockwork view of the world is really much more complicated, as the effects of orbital changes on the seasons is very small, and it is feedback in the climate system that amplifies these changes.

The Causes of the Glacial–Interglacial Cycles

Orbital forcing is insufficient to drive the observed glacial–interglacial variability in climate. Instead, the Earth system amplifies and transforms the changes in solar energy received at the Earth's surface through various feedback mechanisms[5].

Let us start with building an ice age. The first thing that needs to happen is a slight reduction in summer temperatures. As snow and ice accumulate due to this change in summer temperature, the albedo, that is, the reflection of sunlight back into space, increases. The process of reflecting more sunlight back into space suppresses local temperatures, which promotes the accumulation of more snow and ice, which then further increases the albedo of the region, producing the so-called 'ice–albedo' feedback. Once you have a small ice sheet, it changes the local environment, making more snow and ice, and so it can get bigger and bigger.

Another feedback is triggered when the ice sheets, particularly the Laurentide ice sheet on North America, become big enough to deflect the atmospheric planetary waves. This changes the storm path across the North Atlantic Ocean, and prevents the Gulf Stream and North Atlantic Drift from penetrating as far north as it does today. This surface ocean change combined with the general increase in melt-water in the Nordic Seas and Atlantic Ocean due to the presence of a large continental ice sheet ultimately leads to a reduction in the production of deep water. Deep-water production in the Greenland and Labrador Seas is the heartbeat of our modern climate. By reducing the formation of deep water, it reduces the amount of warm water pulled northwards, all of which leads to increased cooling in the Northern Hemisphere and expansion of the ice sheets.

The above physical climate feedbacks are enhanced by complementary changes in atmospheric greenhouse gases. Air bubbles trapped in polar ice have shown us that carbon dioxide dropped by a third, and methane by a half during each glacial period. These changes would have compounded the cooling that occurred during each glacial period, helping to build more ice. The key question is whether Earth's orbital parameters affect the production of greenhouse gases directly, or do changes in the Earth's orbital parameters start ice sheets expansion in the Northern Hemisphere which change global climate and reduce greenhouse gases production. I am sorry to say the jury is still out on this question. What we do know is that greenhouse gases played a critical role in glacial — interglacial cycles. We also know that changes in greenhouse gases concentration always come before changes in global temperatures.

An important question is why these feedbacks do not end up running away and freezing the whole Earth. The answer is that they are prevented from having a runaway effect by a process called 'moisture limitation'. To build an ice sheet, you need it to be cold and wet. However, as the warm surface water is forced further and further south, supply of the moisture that is required to build ice sheets decreases. So, by changing the atmospheric and ocean circulation, the ice sheets end up starving themselves of moisture.

In the last million years, it took up to 80,000 years to build up ice sheets to reach the maximum extent of ice. So, the last time this occurred was about 21,000 years ago. However, getting rid of the ice is much quicker. The process known as 'deglaciation' usually takes a maximum of only 4,000 years. This deglaciation is triggered by an increase in solar energy received during the summer at about 65°N. This encourages the Northern Hemisphere ice sheets to melt slightly. The rise of atmospheric carbon dioxide and methane promotes warming globally and encourages the melting of the large continental ice sheets. These processes have to work against the ice sheets' albedo effect, which produces a microclimate that works to keep them intact. What causes the rapid removal of ice is the rise in sea level due to the melting ice sheet, where large ice sheets adjacent to the oceans are undercut by a rising sea level. The coldest sea water can become is about −1.8°C while the base of the ice sheet is usually colder than −30°C. The effect of the water against the ice is similar to putting hot water under a tub of ice cream. Undercutting of the ice sheet leads to more melting and therefore more caving of ice into the ocean. This in turns increases sea level, which causes even more undercutting. This sea level feedback process can occur extremely rapidly. Once the ice sheets are in full retreat, then the other feedback mechanisms discussed above are thrown into reverse.

Anatomy of the Last Ice Age

If we focus on the last ice age, only 21,000 years ago, we can see what a considerable difference glaciation made to the climate of Earth. In North America, there was nearly continuous ice spreading across the continent from the Pacific to the Atlantic Ocean[6]. It was made up of two separate ice sheets, the Laurentide ice sheet in the east, centred on Hudson Bay, and the Cordilleran ice sheet in the west, in the coastal ranges and the Rockies. The Laurentide covered over 13 million km^2 of land and reached over 3,300 m in thickness at its deepest over the Hudson Bay. Its maximum reach extended from New York, to Cincinnati, to St Louis, and to Calgary. In Europe, there were two major ice sheets, the Fenno-Scandinavian and the British ice sheets, with a minor one over the European Alps. The British ice sheet during many glacial periods merged with the Scandinavian ice sheet. On average, it covered about 340,000 km^2 per glacial period. During the last glacial period, the ice sheet reached halfway down the British Isles to the edge of Norfolk. The Fenno-Scandinavian ice sheet was much larger than the British ice sheet, covering an area of 6.6 million km^2, and extended all the way from Norway to the Ural Mountains in Russia. We must also not forget the Southern Hemisphere as there were significant ice sheets in Patagonia, South Africa, southern Australia, and New Zealand. In addition, the Antarctic ice sheet expanded by about 10% and seasonal sea ice extended an additional 500 miles away from the continent. It is difficult to imagine the huge amount of water that was locked up in these ice sheets. One way of understanding is to consider the oceans. The oceans cover over 70% of our planet, but so much water was sucked out of them and locked up in the ice sheets that the sea level dropped by over 120 m. This is approximately the equivalent of the height of the London Eye. If all the ice on Antarctica and Greenland today melted, it would raise the sea level by 70 m. During a glacial period, global temperatures were 5–6°C lower than today, but this temperature drop was not evenly distributed, with the high latitudes cooling by as much as 12°C, while even the tropics cooled by between 2°C and 5°C. The ice ages were also very dry periods with large amounts of dust in the atmosphere. For example, in Northern China, eastern USA, central and eastern Europe, central Asia, and Patagonia, there are deposits of hundreds of metres of dust called 'loess deposits', which built up during the ice age.

Past Variations in Carbon Dioxide

One of the pieces of scientific evidence that shows that atmospheric CO_2 is an important control of global climate comes from the study of past climate[7]. Evidence for past variations in GHGs and temperature come from ice cores drilled in both Antarctica and Greenland. As snow falls, it is light and fluffy and contains a lot of air. When more snow falls on top, the older snow is slowly compacted to form ice, and some of these air bubbles are trapped. By extracting these air bubbles trapped in the ancient ice, scientists can measure the percentage of GHGs that were present in the past atmosphere. Scientists have drilled down over two miles into both the Greenland and Antarctic ice sheets, which has enabled them to reconstruct the amount of GHGs present in the atmosphere over the past million years. By examining the oxygen and hydrogen isotopes in the frozen

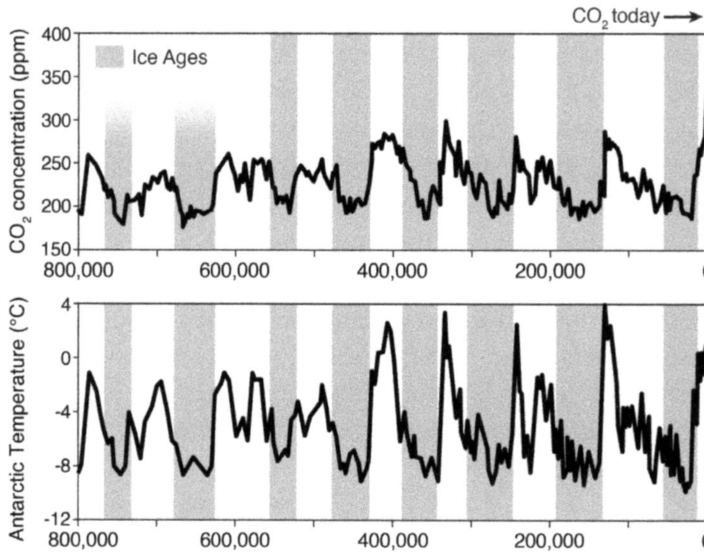

Figure 24. Greenhouse Gases and Temperature for the Past Eight Glacial Cycles Recorded in Ice Cores

water that make up the ice core, it is possible to estimate the air temperature above the ice sheet when the water first froze. The results are striking: the proportion of GHGs such as atmospheric CO_2 and CH_4 co-vary with temperature over the past 800,000 years (see Figure 24). The cyclic changes in climate from glacial to interglacial periods can be seen both in temperatures and in the GHG content of the atmosphere. This strongly supports the idea that GHGs in the atmosphere and global temperature are closely linked; when CO_2 and CH_4 increase, global temperatures increase, and vice versa.

Holocene

We are currently in an interglacial period called the Holocene Epoch, having escaped the last ice age about 10,000 years ago[7]. The climate has not been constant during our interglacial period, and the early Holocene period may have been warmer and wetter than the 20th century. Throughout the Holocene, there have been millennial-scale climate events, Dansgaard-Oescheger cycles, which involve a local cooling of 2°C. These events may have had a significant influence on classical civilisations. For example, the cold arid event about 4,200 years ago coincided with the collapse of many classical civilisations, including the Old Kingdom in Egypt; the Akkadian Empire in Mesopotamia; the Early Bronze Age societies of Anatolia, Greece, and Israel; the Indus valley civilisation in India; the Hilmand civilisation in Afghanistan; and the Hongshan culture of China. The last of these millennial climate cycles was the Little Ice Age. This event is really two cold periods; the first followed the Medieval Warm Period, which ended 1,000 years ago, and is often referred to as the Medieval Cold Period. The Medieval Cold Period played a role in extinguishing Norse colonies on Greenland and also caused famine and mass migration in Europe. It started gradually before AD 1200 and ended at about AD 1650. The second cold period, more classically referred to as the Little Ice Age, may have been the most rapid and greatest change in the North Atlantic region during the late Holocene, as suggested by ice-core and deep-sea sediment records. In Britain, the temperature dropped on average by 1°C — although everyone assumes it was much greater because of the beautiful paintings of the ice fairs on the frozen River Thames. But this is a myth because the alterations to the River Thames means it is now nearly impossible to get the weather in England cold enough to freeze the river. Before the 1830s, the Thames was a slow meandering river. But the demolition of the old London Bridge in 1831, the straightening of the River Thames to build the Embankment in the 1870s for Londoners to promenade like the Parisians, and the dredging of the river to make it an international port at the heart of the British Empire, changed the Thames to a deep straight fast flowing river.

If we examine records around the globe, it seems that the Little Ice Age and the Medieval Warm Period only occurred in northern Europe, northeast America, and Greenland. The Little Ice Age was a regional climate fluctuation driven by small changes in the Gulf Stream and deep-water formation north of Iceland. Many climate change deniers suggest that global warming is just the world recovering from the Little Ice Age. But as most of the world never had a Little Ice Age — there is nothing to recover from. The reconstructed global temperature records for the last two millennia are essential as they provide a context for the instrumental temperature data set for the last 150 years. It is clear from these longer-term records that global temperatures have been much warmer in the 20th and the 21st centuries than at any other time during the last two thousand years.

Early Farmers

The high-resolution ice-core evidence from Greenland and the continental margins of Antarctica shows that GHGs in the atmosphere rose a small amount before the Industrial Revolution in the 1800s. Bill Ruddiman, Professor of Palaeoclimatology at the University of Virginia, suggested that early agriculturalists caused this reversal in natural decline in GHGs[7]. Deforestation and land clearing caused the rise in atmospheric CO_2 starting about 7,000 years ago, while the expansion of wet rice agriculture and cattle caused atmospheric CH_4 to start rising about 5,000 years ago. It seems that early human interactions with our environment increased atmospheric GHGs just enough that even prior to the Industrial Revolution, we had already delayed the onset of the next ice age, which would otherwise have started gently to occur anytime in the next 1,000 years.

The Industrial Revolution

There is clear evidence that levels of atmospheric CO_2 have been rising ever since the beginning of the Industrial Revolution[8]. As mentioned before, the first measurements of CO_2 concentrations in the atmosphere started in 1958, on the summit of Mauna Loa Mountain in Hawaii at an altitude of about 4,000 m. The record clearly shows that atmospheric concentrations of CO_2 have increased every single year since 1958. The mean concentration of approximately 316 parts per million by volume (ppmv) in 1958 has risen to over 420 ppmv today (see Figure 16).

The Mauna Loa observatory CO_2 data can be combined with the detailed ice-core evidence to produce a complete record of atmospheric CO_2 since the beginning of the Industrial Revolution. This shows that atmospheric CO_2 has increased from a pre-industrial concentration of about 280 ppmv to over 420 ppmv at present, representing an increase of over 50%. To put this increase into context, ice-core evidence shows that over the last 800,000 years, the natural change in atmospheric CO_2 has varied between about 200 and 280 ppmv. The variation between warm and cold periods is about 80 ppmv — less than the CO_2 pollution that we have put into the atmosphere over the past 100 years. The level of human pollution caused in one century is greater than natural variations, which took thousands of years.

Anthropocene

There is general scientific agreement that human activity has had a profound geological influence on the Earth System[8]. The magnitude, variety and longevity of human-induced changes to the lithosphere, hydrosphere, cryosphere, biosphere and atmosphere suggest that humans have indeed moved the Earth System beyond the Holocene Epoch. It has been suggested we have entered the Anthropocene Epoch or created the Anthropocene Event.

At the moment the Anthropocene is an informal term used to mark when human actions begin to have a global impact on Earth's environment (Figure 25). It is a very powerful idea that human activity is having geological-scale impacts on Earth, which could last millions of years. It has been argued that humans are having as much effect on the Earth's environment as a giant meteorite, mega-volcanoes or the movement of the continental plates. We could be considered a new geological superpower[9]. Here are the key facts to support this view[8]. Humans move more soil, rock and sediment each year than all the natural processes combined. Humans have created over 170,000 synthetic mineral-like substances, such as plastics, concrete, steel, ceramics and many artificial drugs, compared to the approximately 5,000 'natural' minerals. There are currently 1.4 billion motor vehicles, 2 billion personal computers, and more mobile phones than people on Earth. We make over 300 million tonnes of plastic per year, which is the same weight as everyone on Earth. Our waste plastic can be found in every ocean on our planet. For example, a plastic bag has even been found in the Mariana Trench which is 7 miles deep in the North Pacific Ocean. Microplastics have been found in human blood, tissue and

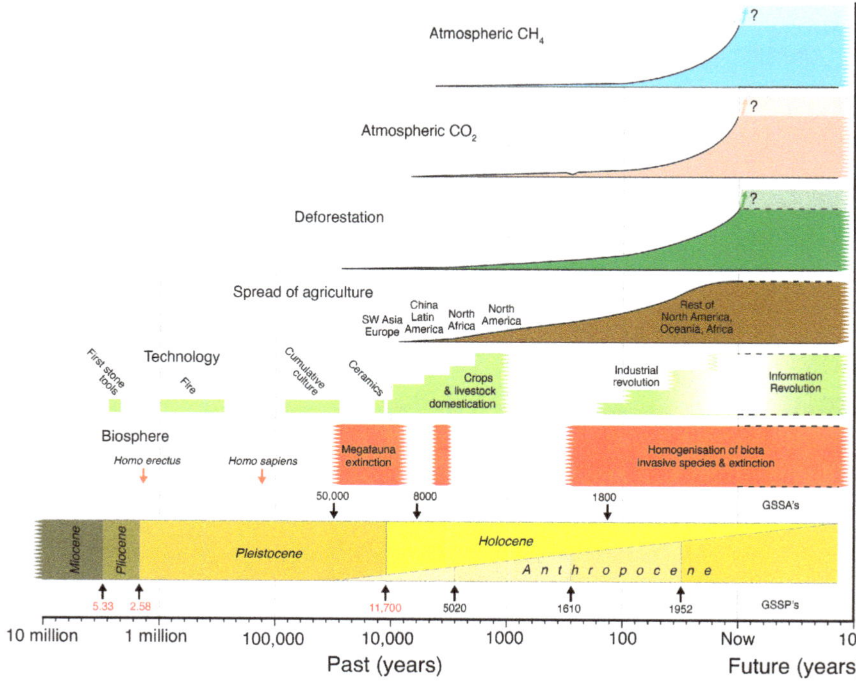

Figure 25. Human Impact Through Time with Potential Start Dates of the Anthropocene Shown

even breast milk. Factories and farms fix more nitrogen in the atmosphere than all naturally occurring processes combined — the last time the nitrogen biogeochemical cycle was this seriously disrupted was over 2 billion years ago when oxygen first started to rise in the atmosphere. Since the beginning of civilisation, we have cut down 3 trillion trees, more than half the trees on Earth. The current weight of all land mammals in the world is made up of 30% humans, 67% livestock and 3% wild animals. 10,000 years ago, wild animals made up 99.95% of the weight of land mammals. We extract 80 million tonnes of fish from the sea, and another 80 million tonnes are farmed every year. Farmland annually produces 4.9 billion heads of livestock and over 4.9 billion tonnes of the top 5 crops: sugar cane, maize, rice, wheat and potatoes. The land-use conversion to produce food, fuel, fibre and fodder, combined with targeted hunting and harvesting, has resulted in species extinctions some 100 to 1,000 times higher than background rates. Since 1500 ACE, there have been 784 documented species extinctions, including 79 mammals, 129 birds, 21 reptiles, 34 amphibians, 81 fish, 359 invertebrates and 86 plants.

Our industry, farming and land-use changes have increased carbon dioxide levels in our atmosphere by over 50%, and methane by over 150% since the beginning of the Industrial Revolution in the mid 18th century. We have added 2.2 trillion tonnes of carbon dioxide to the atmosphere since the beginning of the Industrial Revolution, equivalent in weight to 20 Great Walls of China. Carbon dioxide levels in the atmosphere are now higher than they have ever been in the last 3 million years. Anthropogenic greenhouse gas emissions have also increased the temperature of the Earth by over 1.5°C and caused sea levels to rise by over 25 cm since the beginning of the 20th century.

Summary

The study of past climates is key to understanding how the climate system works and how it might change in the future. We can see how the lithosphere, hydrosphere, cryosphere, biosphere and atmosphere (particularly GHGs) interact with one another on a range of times scales. There is strong evidence that the Earth's climate has cycled from 'greenhouse worlds' to 'icehouse worlds' on a time scale of tens of millions of years. In fact, over the past 50 million years, the Earth's climate has been cooling down, moving towards the cooler and more dynamic 'icehouse world' climate of today. In geological terms, our planet is currently extremely cold, but we are right to worry about global warming. This is because there are huge ice sheets on both Antarctica and Greenland, and nearly permanent sea ice in the Arctic Ocean; this vast amount of ice makes the global climate very sensitive to changes in GHGs. The study of past climate also allows us to compare natural and human rates of change. At the end of the last ice age, the world warmed by 4–5°C in over a period of about 5,000 to 7,000 years. The maximum global warming rate during this transition was about 1.5°C per thousand years. For comparison, humans have warmed the planet by over 1.5°C in the last 100 years — which is approximately 10 times faster than the maximum natural rate of warming.

Chapter 5

Evidence for Climate Change

Science is not a belief system. It is a rational, logical methodology that moves forward by using detailed observation and experiments to constantly test and retest ideas and theories. It is the very foundation of our global society. You cannot just pick and choose which bits of science you want to believe in and which bits you want to reject. For example, you cannot decide that you believe in antibiotics (as they may save your life) or that heavy metal tubes with wings can fly (because you want to go on holiday), and yet at the same time, deny that smoking causes cancer, or that HIV causes AIDS, or that GHGs cause climate change. This chapter presents the scientific evidence that anthropogenic climate change is already happening.

Weight of Evidence

If we are to understand climate change, we must understand how science works. The 'weight of evidence' principle drives the constant need to compile new data and undertake new experiments in order to continually test our ideas and theories regarding climate. Over the past 40 years, the theory of climate change must have been one of the most comprehensively tested ideas in science. There are six main areas of evidence that should be considered[1]:

1. We have tracked the rise in GHGs in the atmosphere and understand their role in past climate variations.

2. We know from laboratory and atmospheric measurements that GHGs do indeed absorb heat when they are present in the atmosphere. Table 1 summarises the latest understanding of the main GHGs.
3. We have tracked significant changes in global temperatures and sea level rise over the last century.
4. We have analysed the effects of natural changes on climate, including sunspots and volcanic eruptions, and though these are essential to understanding the pattern of temperature changes over the past 150 years, they cannot explain warming trend (Figure 17).
5. We have observed significant changes in the Earth climate system, including melting of the Greenland and Western Antarctic ice sheets; retreating of Arctic sea ice; retreating mountain glaciers on all continents; and shrinking permafrost and increased depth of its active layer (the top of the permafrost, which melts every summer).
6. We continually track global weather and have seen significant shifts in the number and intensity of extreme events. These have now been demonstrated as a significant contribution of climate change to many of these extreme weather events.

This chapter will consider evidence for changes in global temperature, precipitation, sea level, and extreme weather events.

What is the IPCC?

The Intergovernmental Panel on Climate Change (IPCC) was established in 1988 jointly by the United Nations Environmental Panel and the World Meteorological Organization to address concerns about the possibility of global warming[2]. The purpose of the IPCC is the continued assessment of the state of knowledge on the various aspects of climate change, including scientific, environmental, and socioeconomic impacts and response strategies. The IPCC does not undertake independent scientific research, rather it brings together all key research published in the world and produces a consensus. There have been six main IPCC Reports, in 1990, 1996, 2001, 2007, 2013/14, and 2021/22 and many individual specialised reports on such subjects as carbon emission scenarios, alternative energy sources, oceans, land use and extreme weather events.

Table 1. Main Greenhouse Gases and their Comparative Ability to Warm the Atmosphere

Greenhouse gas	Chemical formula	Lifetime (years)	Pre-industrial levels	2025 levels	Human source	Global warming potential (compared to CO_2)	
						20 years	**100 years**
Carbon dioxide	CO_2		278 ppmv	430 ppm (>50% increase)	Fossil-fuel combustion Land-use changes Cement production	1	1
Methane	CH_4	12	700 ppbv	1935 ppb (>250% increase)	Fossil-fuels Rice paddies Waste dumps Livestock	83	30
Nitrous oxide	N_2O	109	275 ppbv	339 ppb (>20% increase)	Fertilizer Industrial processes Fossil-fuel combustion	273	273
CFC-12	CCl_2F_2	100	Not naturally occurring	~500 ppt	Liquid coolants/foams	10,800	10,200
HCFC-22	$CHClF_2$	12	Not naturally occurring	~240 ppt	Liquid coolants	5,280	1,760
Perfluoro methane PCF-14	CF_4	50,000	0*	~85 ppt	Production of aluminium	5,301	7,380
Sulphur hexa-fluoride	SF6	3,200	0*	~12 ppt	Dielectric fluid	17,500	23,500

ppm = parts per million in the atmosphere
ppb = parts per billion in the atmosphere
ppt = parts per trillion in the atmosphere
*trace amounts are found naturally

The IPCC is recognised as the most authoritative scientific and technical voice on climate change, and its assessments have had a profound influence on the negotiators of the United Nations Framework Convention on Climate Change (UNFCCC). The IPCC is organised into three working groups plus a task force to calculate the amount of GHGs produced by each country. Each of these three bodies have two co-chairs (one from a developed and one from a developing country) and a technical support unit. Working Group I assesses the scientific aspects of the climate system and climate change; Working Group II addresses the vulnerability of human and natural systems to climate change, and options for adapting to climate change; and Working Group III assesses options for limiting GHGs emissions and otherwise mitigating climate change. The IPCC provides governments with scientific, technical, and socioeconomic information relevant to evaluating the risks and to developing a response to global climate change[2]. The latest reports from these three working groups were published in 2022, with approximately 500 experts, from some 120 countries, directly involved in drafting, revising, and finalising the IPCC reports, as well as thousands of additional experts participating in the review process. The IPCC authors are always nominated by governments and international organisations, including non-governmental organisations[3]. These reports are essential reading for anyone interested in global warming and are listed in the Further reading section at the end of the book. In 2008, the IPCC was jointly awarded, with Al Gore, the Nobel Peace Prize, to acknowledge all the work the IPCC had done over the previous 20 years.

Temperature

Temperatures can be estimated from a number of sources, both direct thermometer-based and proxy-based indicators. Proxy-based indictors are variables that are measured when direct measurements are not available or not possible. For example, infrared (heat) satellite measurements are proxy-based indicators that can be used to estimate surface temperatures.

Direct thermometer-based measurements of air temperature have been recorded at a number of sites in North America and Europe from as far back as 1760[4]. The number of observation sites did not increase to sufficient worldwide geographical coverage to permit a global land average to be calculated until about the middle of the 19th century. Sea-surface temperatures (SSTs) and marine air temperatures (MATs) were systematically recorded by ships from the mid-19th century, but even today the coverage of the Southern Hemisphere is extremely poor. All these data sets require various corrections to account for changing conditions and measurement techniques. For example, each land data station has been examined to ensure that conditions have not varied through time as a result of changes in the measurement site, instruments used, instrument shelters, or the way monthly averages were computed. The growth of cities around some of the measuring sites, which leads to warmer temperatures caused by the urban heat island effect, must also be accounted for. In the IPCC science report, the influence of the urban heat island effect is acknowledged as real, but even if it were not accounted for, it would still have a negligible effect for the global temperature compilation (less than 0.006°C).

For SST and MAT, there are a number of corrections that have to be applied[5]. First, up to 1941, most SST temperature measurements were made in seawater hoisted on deck in a bucket. Since 1941, most measurements have been made at the ships' engine water intakes. Second, between 1856 and 1910, there was a shift from wooden to canvas buckets, which changes the amount of cooling caused by evaporation that occurs as the water is being hoisted on deck. In addition, through this period, there was a gradual shift from sailing ships to steamships, which altered the height of the ship decks and the speed of the ships, both of which can affect the evaporative cooling of the buckets. The other key correction that has to be made is for the global distribution of meteorological stations through time, which has varied greatly since 1870.

The collation of the global temperature records has been undertaken by a number of groups around the world, including the UK Met Office, NASA, National Oceanic and Atmospheric Administration (NOAA) and the Japan Meteorological Agency (see Figure 26). In 2012, Richard Muller, a physicist (who was then a climate change sceptic), and his Berkeley group collated global temperature records for the last 250 years. As his group had not taken into account of all the corrections mentioned above, their estimation of global warming was larger than the other

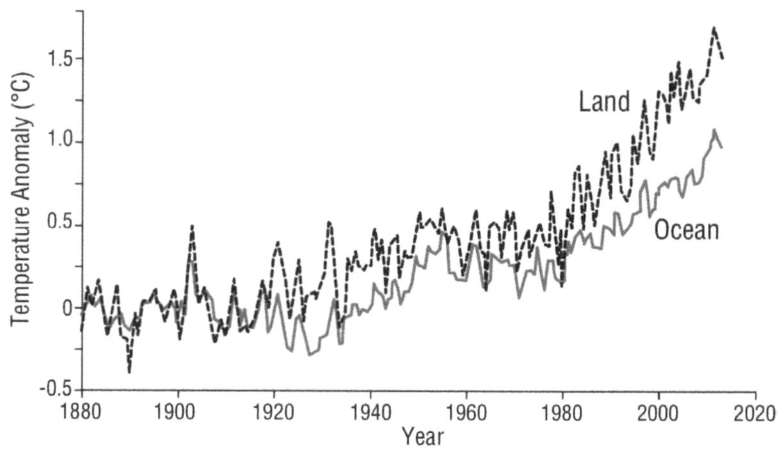

Figure 26. Land and Ocean Temperature Data Since 1850

groups. This was subsequently revised, and Richard Muller publicly announced he had changed his mind about climate change. He stated that climate change was occurring and was clearly due to human activity — and now, the Berkeley group temperature records are widely used[6].

By making all the necessary corrections, it is possible to produce a continuous record of global surface temperature from 1880 to 2025, which shows an observed warming of over 1.5°C over this period (Figure 26). These observations are supported by 60 years of balloon and satellite data. For example, there are over 800 stations that twice a day release rawinsondes (meteorological instruments), or balloons, to measure temperature, relative humidity, and pressure through the atmosphere to a height of about 20 km, where they burst. The temperature records also show us that the land is warming up faster than the oceans (Figure 26).

Global temperatures have also been reconstructed for periods of time pre-dating instrumental or thermometer records. This has been achieved by using paleoclimate proxies such as the thickness of tree rings and the isotopic composition of ice cores or cave deposits to estimate local temperatures. A study published in *Nature* in 2019 led by Raphael Neukom at the Oeschger Centre for Climate Change Research (University of Bern, Switzerland) used over 700 paleoclimate records and showed that in the last 2,000 years[7], the only time the climate all around the World has changed at

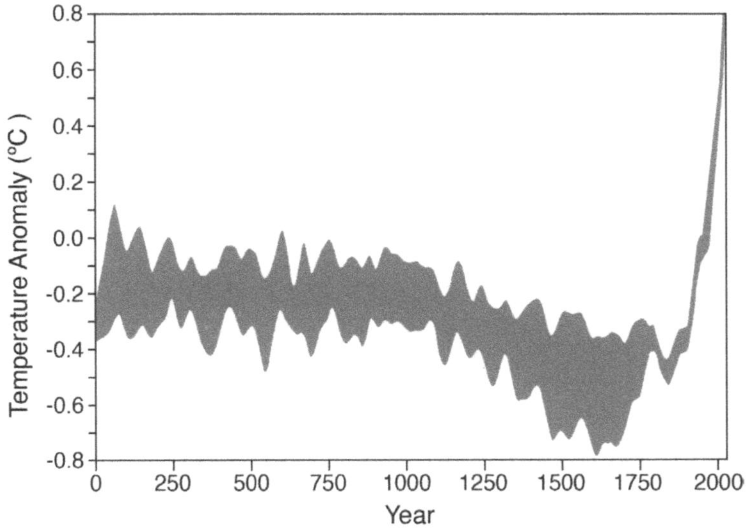

Figure 27. Northern Hemisphere Temperature Reconstruction for the Last 2000 Years[7]

the same time and in the same direction has been in the last 150 years, when over 98% of the surface of the planet has warmed (see Figure 27).

Precipitation

There are two global precipitation data sets: Hulme and the Global Historical Climate Network (GHCN)[4]. Unfortunately, unlike temperature, rainfall and snow data are not as well documented and recording has not been carried out for as long. It is also known that precipitation overland tends to be underestimated by up to 10–15% owing to the effects of air-flow around the collecting dish. Without correction of this effect, a spurious upward trend could be perceived in global precipitation. Despite these problems, there seems to be a significant increase of precipitation over the past 25 years (see Figure 28), particularly in the Northern Hemisphere middle latitudes. This is supported by evidence that since the 1980s, atmospheric water content has increased over the land and ocean as well as in the upper troposphere. This is consistent with the extra water vapour that the warmer atmosphere can hold.

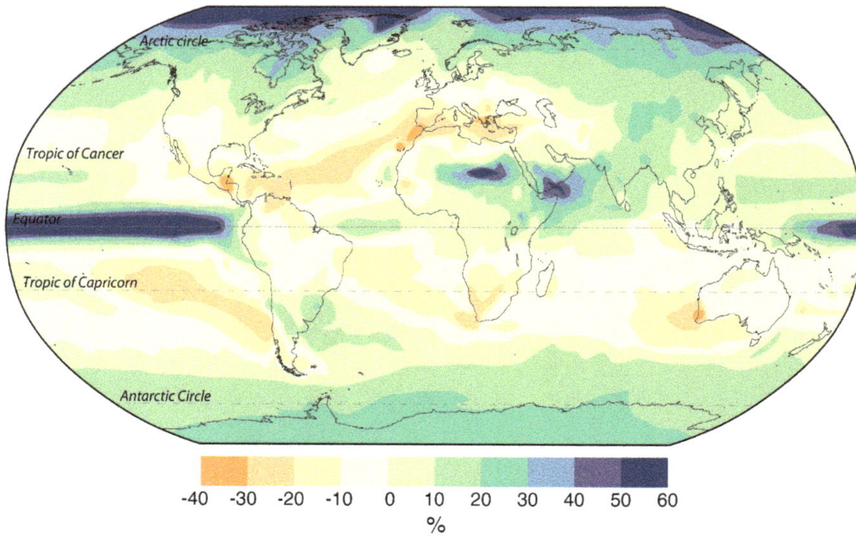

Figure 28. Global Precipitation Changes (1900–2018)[1]

There is evidence for a global increase in precipitation but the evidence for this change is much stronger when considering individual regions. The latest IPCC report[1] suggests that significant increases in precipitation have occurred in the eastern parts of North and South America, northern Europe, as well as northern and central Asia. It seems that seasonality of precipitation is also changing, for example, in the high latitudes in the Northern Hemisphere, with increased rainfall in the winter and a decrease in the summer. Long-term drying trends have been observed on the Sahel, in the Mediterranean, southern Africa, and parts of southern Asia. It has also been observed that the amount of rain falling during heavy 'extreme' rain events has increased.

Relative Global Sea Level

Records show that between 1901 and 2025, the global sea level has risen by over 25 cm (see Figure 29)[1]. Sea-level change is difficult to measure, as relative sea-level changes have been derived from two very different data sets — tide-gauges and satellites. In the conventional tide-gauge

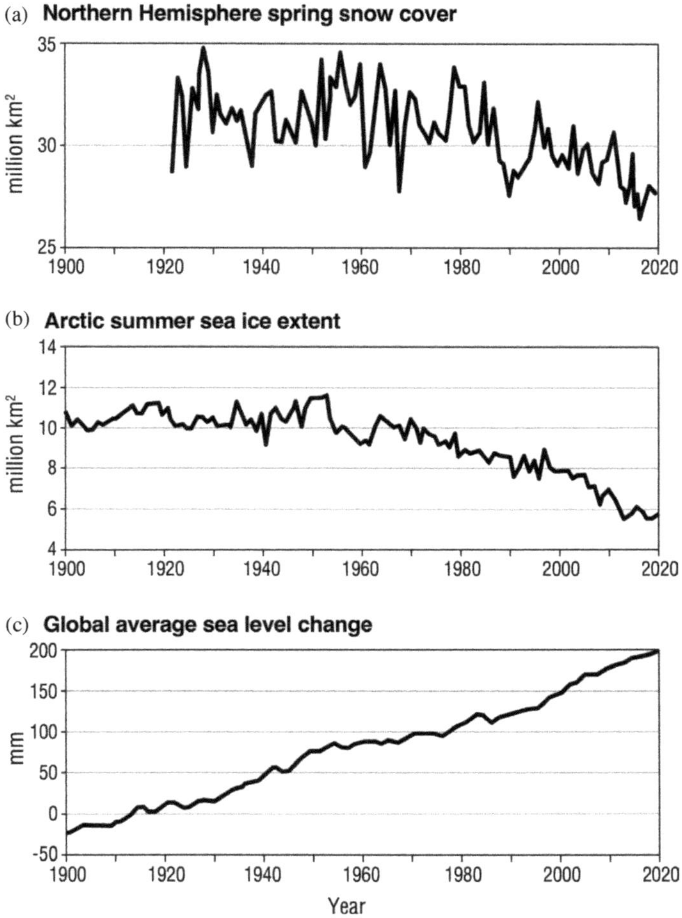

Figure 29. Indicators of Climate Change[1]

system, the sea level is measured relative to a land-based tide-gauge benchmark. The major problem is that the land surface is much more dynamic than one would expect, with a lot of vertical movements, and these become incorporated into the measurements. Vertical movements can occur as a result of normal geological compaction of delta sediments, the withdrawal of groundwater from coastal aquifers, uplift associated with colliding tectonic plates (the most extreme of which is mountain-building such as in the Himalayas), or ongoing post-glacial rebound, and compensation elsewhere, associated with the end of the last ice age. The rebound is caused by the rapid removal of weight when the giant ice sheets melted, so that the land that has been weighed down slowly rises back to its original position. An example of this is Scotland, which is rising at a rate of 3 millimetres (mm) per year, while England is still sinking at a rate of 2 mm per year, despite the Scottish ice sheet having melted 10,000 years ago. In comparison, the problem with the satellite data is that the time covered is too short. The best satellite data started in January 1993 and shows a trend of over 35 mm rise in sea level per decade. This means satellite data have to be combined with the tide-gauge data to look at long-term trends.

In summary, between 1900 and 2025, the global average sea level rose by about 2 mm per year; with the fastest rise in sea level observed in the last decade at 4.62 mm per year — over twice the 20th century rate[1]. The sea-level rise of the last 30 years is made up of the following contributions: 39% from thermal expansion of the ocean; 9% Antarctic ice sheet; ~12% Greenland ice sheet; 27% glaciers and other ice caps; and another ~13% due to the overall reduced land storage of water. The Greenland and Antarctic ice sheets have contributed to recent sea-level rise and this contribution is accelerating. At the moment, it is estimated that Greenland is losing over 230 gigatonnes of ice per year, a seven-fold increase since the early 1990s. Meanwhile, Antarctica is losing about 150 gigatonnes of ice per year, a five-fold increase since the early 1990s, and most of this loss is from the northern Antarctic Peninsula and the Amundsen Sea sector of West Antarctica (Figure 30).

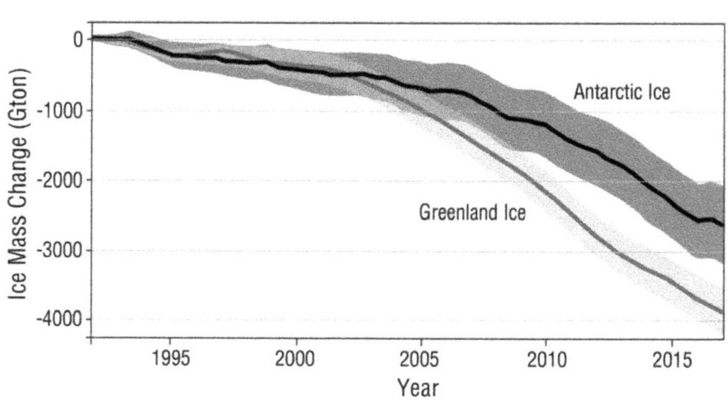

Figure 30. Melting of Antarctic and Greenland Ice Sheets

Other Evidence for Global Warming

Other evidence for climate change comes from the high latitudes and from monitoring extreme weather events[1]. The annual mean Arctic sea ice extent has decreased in total since 1979 at a rate of approximately 4% per decade, which means a loss of half a million km^2 per decade. The summer sea ice minimum has decreased even more by 12.8% per decade, which is equivalent to a loss of 1 million km^2 per decade. Since 1979, the annual mean Antarctic sea ice extent has varied markedly with record highs and lows, but over the last decade, there has been a significant decline. 2025 was the fourth consecutive year when the minimum Antarctic sea ice extent was below 2.0 million square kilometres, confirming a 2.3% per decade loss of sea ice.

There is also evidence from permafrost regions. Permafrost exists in high-latitude and high-altitude areas, where it is so cold that the ground is frozen solid to a great depth. During the summer months, only the top half metre or so of the permafrost becomes warm enough to melt, and this is called the 'active layer'. There has been a 3°C warming in Alaska and 2°C warming in northern European/Russia permafrost over the last 50 years, and evidence shows that the active layer has become much deeper. The maximum area covered by seasonal permafrost has decreased by 7% in the Northern Hemisphere since 1900, with a decrease in the spring of up to 15%. This increasingly dynamic cryosphere will amplify the natural hazards for people, structures, and communication links. Already we have seen this in the form of damage to buildings, roads, and pipelines, such as the oil pipelines in Alaska. In addition, there is evidence that most, if not all non-ice-sheet glaciers are in retreat. The amount of total snowfall and the annual snow and ice cover, particularly in the Northern Hemisphere, has greatly reduced (Figure 29). Since 1922, over 0.27 million km^2 of snow and ice cover per decade has been lost. In the Arctic, snow-cover duration has decreased on average by 5 days per decade and larger declines have occurred in the Eurasian Arctic region (~12.6 days) and North American Arctic region (6.2 days).

There is also evidence that spring is occurring earlier in the Northern Hemisphere. The ice cover records from the Tornio River in Finland, which have been compiled since 1693, show that the spring thaw of the frozen river now occurs a month earlier. In Kyoto, Japan, the famous cherry blossoms now appear 11 days earlier than 100 years ago[8], and the earliest blooming recorded in the whole of the 1200-year long record was in 2021. In South Korea, some species are appearing over 20 days earlier compared to a century ago, including studies plum, forsythia, and cherry blossoms. In France, the grape harvest in Beaune is now 10 days earlier than 100 years ago. In Britain, among the variety of indicators of an earlier spring is evidence of birds nesting over 12 days earlier than 45 years ago. Meanwhile, in the US, species that are active in early spring such as lilac and honeysuckle are unfurling their leaves three weeks earlier than 40 years ago.

Extreme Weather Events

The latest IPCC report states that it is virtually certain that anthropogenic climate change has caused increases in the frequency and severity of hot extremes and decreases in cold extremes on most continents. The frequency and intensity of extreme weather events such as storms, floods, heatwaves, droughts and wildfires have increased on every continent. In 2024, there were over 600 extreme weather events recorded of which over 150 were unprecedent — meaning the intensity, duration or location was record breaking[9].

Human climate change has also played a role in shaping the global distribution and intensity of tropical cyclones. A 2020 study by James P. Kossin at NOAA and colleagues[10] showed a 15% increase in the occurrence of the most destructive cyclones around the world over the past 40 years. Most marked was the 49% per decade increase in major hurricanes occurring in the North Atlantic and the 18% per decade increase in major cyclones in the southern Indian Ocean. In summary, the number of tropical cyclones originating in the North Atlantic Ocean, Pacific Ocean and southern Indian Ocean has increased as well as the year-to-year variability. For example, in 2019, there were four spectacular cyclones in the Indian Ocean and two of these, in the southern Indian Ocean, were unprecedented. The 2018-19 South-West Indian Ocean cyclone season was the costliest and most active season ever recorded since reliable records began in 1967. In 2020, Super Cyclonic Storm Amphan formed in the North Indian Ocean and made landfall in west Bengal, affecting nearly 40 million people and causing over $13 billions in damages.

The most powerful tropical storm up to now was Typhoon Haiyan[11] (known in the Philippines as Super Typhoon Yolanda) in 2013. It had sustained winds reaching 195 mph (314 km/h). Over 6000 people died, 29,000 people were injured, and 14.1 million people were affected, of which 4.1 million people were made homeless. The overall cost of damage was around $12 billion with 1.1 million tonnes of crops destroyed, 1.1 million houses damaged and over 1 million farmers and 600,000 hectares of farmland affected.

The reason why scientists are certain that many of these extreme weather events are made worse by climate change is the field of attribution science. Advances in computer processing power and improved methods for modelling the factors that contribute to weather allow scientists to run weather simulations for a region with and without anthropogenic GHG forces. This allows us to determine the extent to which climate change has contributed to specific extreme weather events, and if there has been a contribution, whether it has increased the intensity or the frequency or both. Over 750 extreme weather events that occurred in the last two decades have been studied using attribution science (Figure 31). 74% of the studied events were found to have increased frequency or intensity due to climate change, with some events being virtually impossible without it[10].

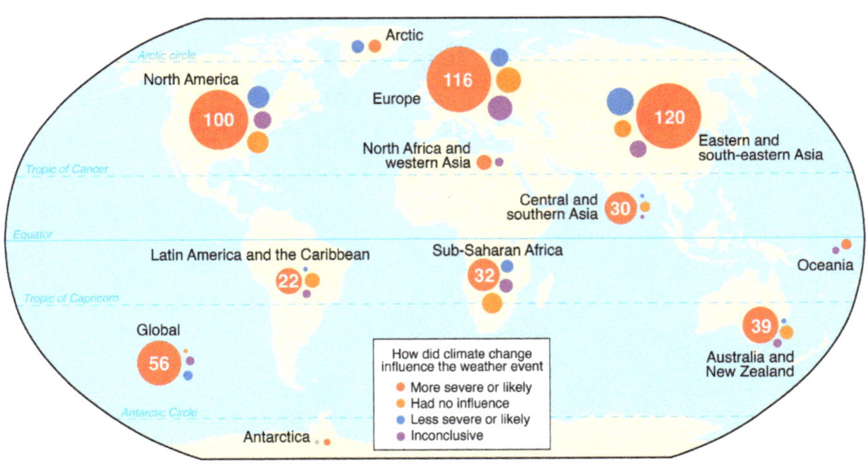

Figure 31. Studies of Almost 750 Events and Trends Reveal the Impact of Climate Change on Extreme Weather[10].

Summary

Over the past 150 years, significant changes in climate have been recorded, changes which are markedly different from those of at least the past 2,000 years. These changes include over 1.5°C increase in average global temperatures, sea-level rise of over 25 cm, significant shifts in the seasonality and intensities of precipitation, changing weather patterns, the accelerated melting of the Greenland and Western Antarctic ice sheets, as well as the significant retreat of Arctic sea ice and nearly all continental glaciers. According to the US NOAA, between 1880 and 2026, the 11 warmest years on record have all occurred within the past 11 years, with 2024 the warmest year on record[12]. The IPCC 2021 science report[1] states that the evidence for climate change is unequivocal, and there is very high confidence that this warming is due to human emissions of GHGs. This statement is supported by six main lines of evidence: (1) the rise in GHGs in the atmosphere has been measured and the isotopic composition of the gases shows that the majority of the additional carbon comes from the burning of fossil fuels; (2) laboratory and atmospheric measurements show that these gases absorb heat; (3) significant changes in global temperatures and sea-level rise have been observed over the past century; (4) other significant changes have been observed in the cryosphere, oceans, land, and atmosphere including retreating ice sheets, sea ice and glaciers, as well as extreme weather events which can all be directly attributed to the impact of climate change; (5) there is clear evidence that natural processes including sun spots and volcanic eruptions cannot explain the warming trend over the past 100 years; and (6) we now have a deeper understanding of the longer term climate changes of the past and the critical role GHGs have played in regulating the climate of our planet.

Chapter 6

Modelling Future Climate

The whole of human society operates on knowing the future weather. For example, farmers in India know when the monsoon rains will come next year and so they know when to plant the crops. Farmers in Indonesia know there are two monsoon rains each year, so they can have two harvests yearly. This is based on their knowledge of the past, as the monsoons have always come at about the same time each year in living memory. But the need to predict goes deeper than this; it influences every part of our lives. Our houses, roads, railways, airports, offices, cars, trains, and so on are all designed for the local climate. Predicting future climate is now essential, as we can no longer rely on records of past weather to tell us what the future will hold. We also need to understand the consequences of our actions. For example, if we continue to emit GHGs at the same rate as today, how much climate change will occur? New ways have been developed to understanding potential futures — We model the future.

Models

There is a whole hierarchy of climate models[1], from relatively simple box models to extremely complex three-dimensional General Circulation Models (GCMs). Each of them has a role in examining and furthering our understanding of the global climate system. It is the complex three-dimensional GCMs that are used to predict future global climate. These comprehensive climate models are based on physical laws represented by mathematical equations, which are solved using a three-dimensional grid over the globe. To obtain the most realistic simulations, all the major parts of the climate system must be represented in sub-models, including atmosphere, ocean, land surface (topography), cryosphere, and biosphere, as well as the processes that go on within them and between them.

Over the past 40 years, there has been a huge improvement in climate models. This has been largely due to our increased knowledge of the climate system but also because of the nearly exponential growth in computer power. There has been a massive improvement in spatial resolution of the models[2] from the very first IPCC report in 1990 to the latest in 2021. The current generation of GCMs have multiple layers in the atmosphere, land, and ocean, as well as a spatial resolution greater than one point every 30 km by 30 km (see Figure 32). Equations are typically solved for every simulated 'half-hour' of a model run. Many physical processes, such as atmospheric chemistry, the formation of clouds, production and movement of aerosols (particles suspended in the air), and ocean convection, take place on a much smaller scale than the main model can resolve. The effects of small-scale processes have to be lumped together, which is referred to as 'parameterisation'. All of these parameterisations are checked with separate 'small-scale-process models' to validate the scaling up of these smaller influences[3].

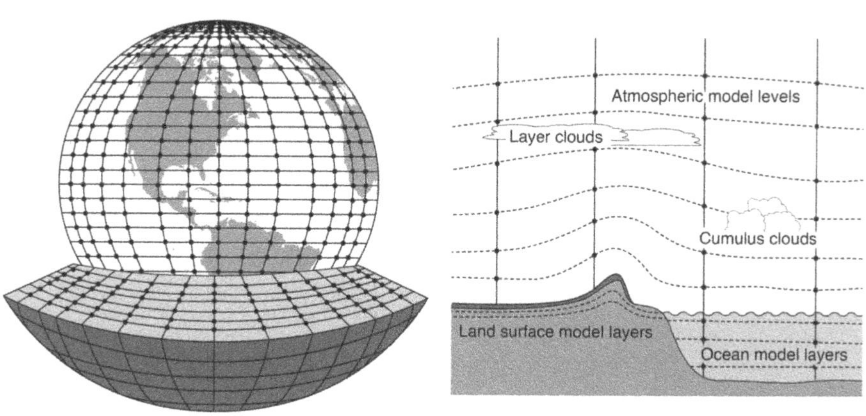

Figure 32. Generic Structure of a Global Climate Model

The biggest unknown in the models is not the physics or the chemistry or the biology; it is the estimation of future global GHG emissions over the next 80 years. This includes many variables, from the global economy to personal lifestyles. Individual models are therefore run many times with different emission scenarios to provide a range of changes that could occur in the future. In fact, the latest IPCC sixth assessment (AR6) report[2] has complied the results of multiple runs from over 100 distinct climate models being produced across 49 different international modelling groups, which are all part of the Coupled Model Intercomparison Projects (CMIP6). Of course, as computer processing power continues to increase, both the representation of coupled climate systems and the spatial scale will likewise improve.

Carbon Cycle

At the heart of the climate models is the carbon cycle, central to estimating what happens to anthropogenic carbon dioxide and methane emissions. The Earth's carbon cycle is complicated, with both large sources and sinks of carbon dioxide. Currently, half of all our carbon emissions are absorbed by the natural carbon cycle and do not end up in the atmosphere, but rather in the oceans and the terrestrial biosphere[4]. Figure 33 shows the global carbon reservoirs in gigatonnes (GtC, or 1,000 million tonnes) and fluxes (the ins and outs of carbon in GtC per year). These indicative figures show the changes since the Industrial Revolution. Evidence is accumulating that many of the fluxes can vary significantly from year to year.

This is because in contrast to the static view conveyed in figures like this one, the carbon system is dynamic, and coupled with the climate system on seasonal, inter-annual, and decadal timescales. What has become clear is that the ocean surface and the land biosphere both take up about 25% each of our carbon emission every year. As the oceans continue to warm, they can hold less dissolved carbon dioxide, which means their uptake will reduce. As we continue to deforest and substantially alter land use, the ability of the land biosphere to absorb carbon diminishes.

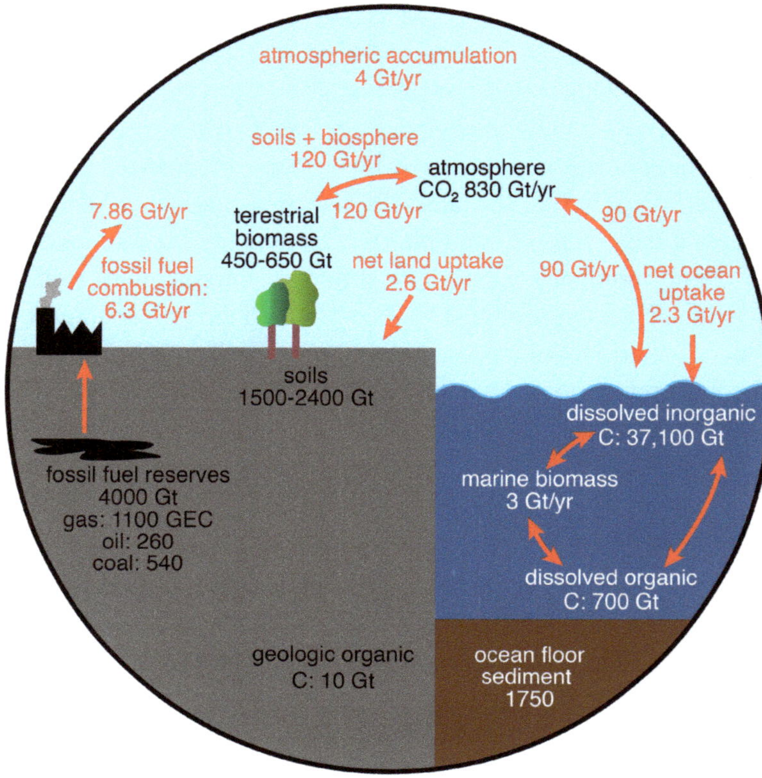

Figure 33. The Carbon Cycle, in Gigatonnes of Carbon (GtC)

Warming and Cooling Effects

To complicate matters, the Earth's climate system also has cooling effects in opposition to the warming effects of the GHGs (see Figure 34)[5]. This includes the amounts of aerosols in the air (many of which come from human pollution, such as sulfur emissions from power stations) and these have a direct effect on the proportion of solar radiation that hits the Earth's surface. Aerosols have a significant local or regional impact on temperature. Computer simulations of climate change demonstrate that industrial areas of the planet have not warmed as much as predicted from rising GHGs. This so called 'global dimming', or more precisely, 'regional dimming', has been confirmed with real temperature and aerosol measurements.

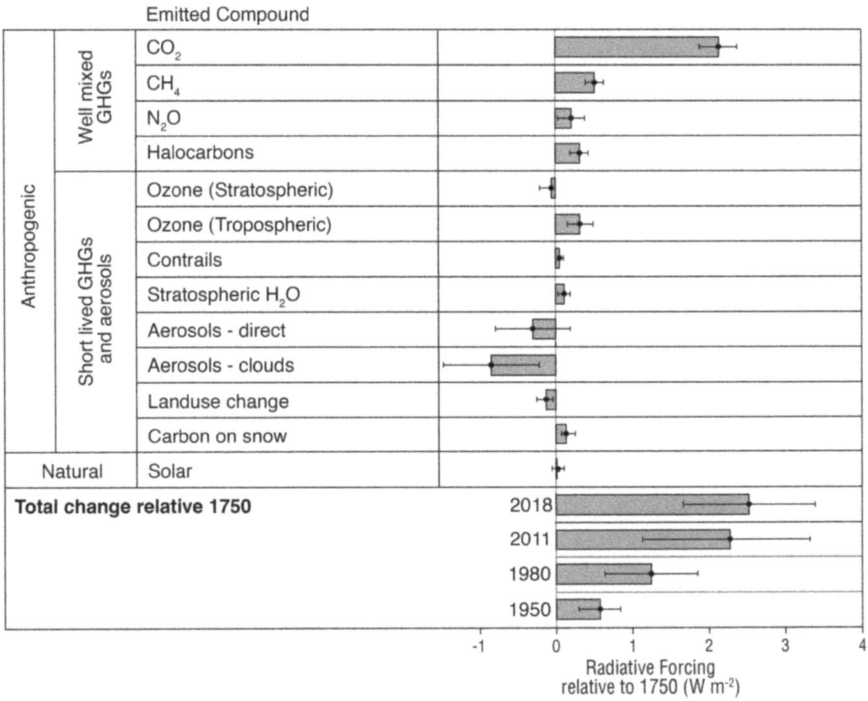

Figure 34. Radiative Forcings between 1750 and 2018[2]

Water vapour is a GHG, but, at the same time, the upper white surface of clouds reflects solar radiation back into space. The level of reflectivity of a surface is called 'albedo'. Clouds and ice have a high albedo, which means that they reflect large quantities of solar radiation away from surfaces on Earth. Increasing aerosols in the atmosphere increases the amount of cloud cover, as they provide points on which the water vapour can nucleate. Predicting what will happen to the amount and types of clouds, and their warming or cooling potential, has been one of the key challenges for climate scientists.

There are also other natural forcings that can cause cooling and warming, and these include sunspot cycles and volcanic eruptions. Evidence shows that volcanic eruptions can have a local transient cooling effect. For example, the mega eruption of Mount Tambora in modern-day Indonesia in April 1815 only caused the average global temperature to decrease by

0.4–0.7 °C for a year. This led to the year 1816 being known as the 'Year Without a Summer'. In 1883, Krakatoa erupted, killing 36,417 people. The eruption is considered to be the loudest sound ever heard in modern history, with reports of it being heard nearly 3,000 miles away. It was equivalent to 200 megatons of TNT, which is about 13,000 times the nuclear yield of the Little Boy bomb that devastated Hiroshima, Japan, during World War II. The sulphur dioxide and dust injected into the atmosphere increased the amount of sunlight reflected back into space, and average global temperatures fell by as much as 1.2°C in the year following the eruption. But in less than 5 years, the weather patterns and temperatures had return to normal.

There is also evidence that the 11-year solar cycle, during which the Sun's energy output varies by roughly 0.1%, can influence ozone concentrations, temperatures, and winds in the stratosphere. These changes, however, have only a very small effect on surface temperatures. Figure 35 shows that since 1880, the solar radiation increased gradually to a peak in about 1955, and since then, it has been decreasing. Over the past 50 years, when global temperatures increased dramatically, solar output has in fact decreased[6].

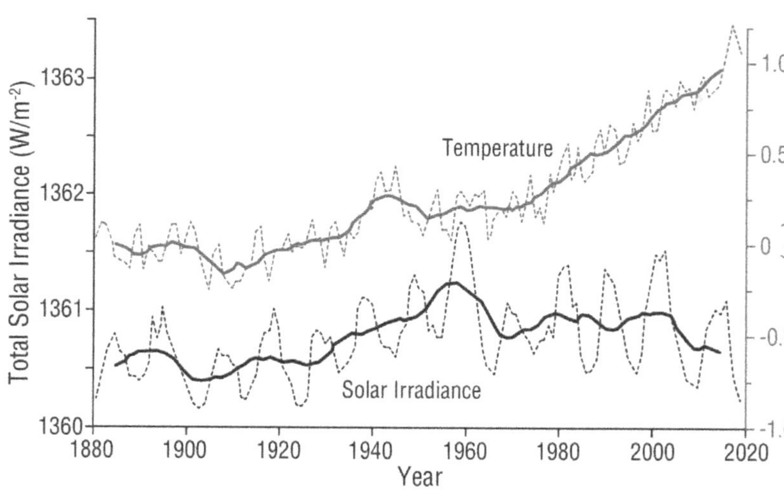

Figure 35. Sunspot and Global Temperatures

Emission Models of the Future

As noted earlier, a critical problem with trying to predict future climate is predicting the amount of carbon dioxide emissions that will be produced in the future. This is influenced by population growth, economic growth, development, fossil-fuel usage, the rate at which we switch to alternative energy, the rate of deforestation, and the effectiveness of international agreements to cut emissions. Out of all the systems that we are trying to model into the future, humanity is by far the most complicated and unpredictable. If you want to understand the problem of predicting what will happen in the next 75 years, imagine yourself in 1920 and what you would have predicted the world to be like in the 21st century. At the beginning of the 20th century, the British Empire was the dominant world power due to the Industrial Revolution and the use of coal. Would you, at that time, have predicted the switch to a global economy based on oil after the Second World War? Or the explosion of car use? Or the general availability of air travel? Even 30 years ago, it would have been difficult to imagine that there would be budget airlines, allowing for cheap flights throughout Asia, Europe, and USA.

The first IPCC reports used simplistic assumptions of GHG emissions over the next 100 years. From 2000 onwards, the climate models used more detailed scenarios set down in an IPCC special report. The 2013 IPCC Fifth Assessment report used more sophisticated Representative Concentration Pathways (RCPs), which considered a much wider variable input to the socio-economic models, including population, land use, energy intensity, energy use, and regional differentiated development. The RCPs were defined by the final radiative forcing achieved by the year 2100, and they range from 2.6 to 8.5 watts per square metre (W/m^2). Radiative forcing is defined as the difference between sunlight (radiant energy) received by the Earth and the energy radiated back to space and is measured in units of W/m^2 of the Earth's surface. For the 2021 IPCC Sixth Assessment report, the RCP1.9 was added (Figure 36), to represent the target to limit global temperatures to 1.5°C above pre-industrial levels at the COP21 Paris 2015 climate conference (see Chapter 10 for full details of this agreement).

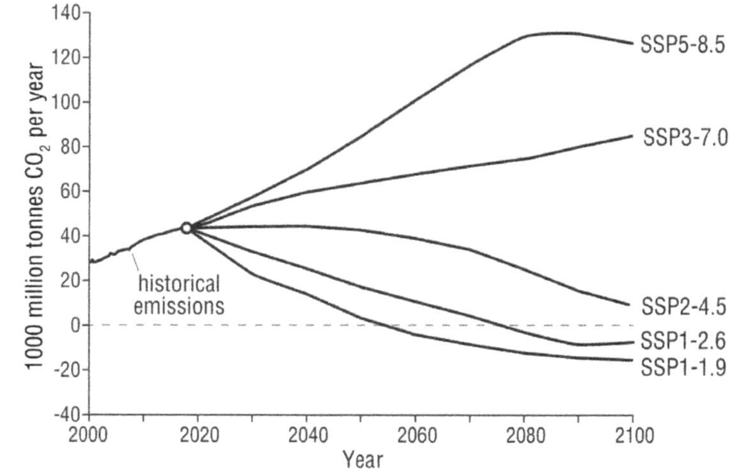

Figure 36. Future Carbon Emission Scenarios[7]

The IPCC Sixth Assessment report also uses Shared Socioeconomic Pathways (SSPs), which were developed to cover the full range of possible futures. The SSPs are a set of five narratives and driving forces that may shape the global economy and global emissions in the future. The SSPs were defined in 2017 by Keywan Riahi and colleagues in the journal *Global Environmental Change* as described below[7].

SSP1: Sustainability — Taking the Green Road

Low challenges to mitigation and adaptation: In this scenario, the world shifts gradually but continually towards a more sustainable path, with inclusive and environmentally aware economic development. Management of the global commons slowly improves; educational and health investments accelerate the demographic transition, reducing population, and the emphasis on economic growth shifts towards a broader emphasis on human wellbeing. Driven by an increasing commitment to achieving development goals, inequality is reduced both across and within countries. Consumption is oriented towards low material growth and lower resource and energy intensity.

SSP2: Middle of the Road

Medium challenges to mitigation and adaptation: In this scenario, the world follows a path in which social, economic, and technological trends do not shift markedly from historical patterns. Development and income growth proceeds unevenly, with some countries making relatively good progress while others fall short of expectations. Global and national institutions work toward, but make slow progress in achieving sustainable development goals. Environmental systems experience degradation, although there are some improvements, and overall, the intensity of resource and energy use declines. Global population growth is moderate and levels off in the second half of the century. Income inequality persists or improves only slowly and challenges to reducing vulnerability to societal and environmental changes remain.

SSP3: Regional Rivalry — A Rocky Road

High challenges to mitigation and adaptation: In this scenario, there is a resurgence of nationalism, concerns about competitiveness and security, as well as regional conflicts that push countries to increasingly focus on domestic or, at most, regional issues. Policies shift over time to become increasingly oriented toward national and regional security issues. Countries focus on achieving energy and food security goals within their own regions at the expense of broader-based development. The scenario also assumes that investments in education and technological development decline. Economic development is slow, consumption is material-intensive, and inequalities persist or worsen over time. Population growth is low in industrialised nations and high in developing countries. A low international priority for addressing environmental concerns leads to strong environmental degradation in some regions.

SSP4: Inequality — A Road Divided

Low challenges to mitigation, but high challenges to adaptation: In this scenario, there are highly unequal investments in human capital, combined with increasing disparities in economic opportunity and political power, leading to increasing inequalities and stratification both across and within countries. Over time, a gap widens between an internationally-connected society that contributes to knowledge and capital-intensive sectors of the global economy and a fragmented collection of lower-income, poorly educated societies that work in a labour intensive, low-tech economy. Social cohesion degrades, and conflict and unrest become increasingly common. Development is substantial only in the high-tech economies. The globally connected energy sector diversifies, with investments in both carbon-intensive fuels like coal and unconventional oil, as well as low-carbon energy sources. Environmental policies focus on local issues in the middle and high-income areas.

SSP5: Fossil-Fuelled Development — Taking the Highway

High challenges to mitigation, but low challenges to adaptation: In this scenario, the world places increasing faith in competitive markets and innovation, with participatory societies producing rapid technological progress and development of human capital as the path to sustainable development. Global markets are increasingly integrated. There are also strong investments in health, education, and institutions to enhance human and social capital. At the same time, the push for economic and social development is coupled with the exploitation of abundant fossil fuel resources and the adoption of resource and energy-intensive lifestyles around the world. All these factors lead to rapid growth of the global economy, while global population peaks and declines in the 21st century. Local environmental problems like air pollution are successfully managed. There is faith in the ability to effectively manage social and ecological systems, including by geoengineering if necessary.

These are narratives that describe the different pathways our future society could take. SSP1 and SSP5 are optimistic about human development, with both allowing for "substantial investments in education and health, rapid economic growth, and well-functioning institutions". The difference is that SSP5 is fossil fuel energy intensive, while SSP1 assumes a shift to renewable energy. SSP3 and SSP4 are pessimistic about the future and SSP2, as it says, is a middle-of-the-road scenario. The scenarios used in the Sixth Assessment report are a combination of the SSPs and RCPs,

Table 2. Defining Representative Concentration Pathway

Representative Concentration Pathway (RCP)	Description
RCP8.5	Rising radiative forcing pathway leading to 8.5 W/m^2 (~1000 ppm CO_2 between 2081 and 2100)
RCP7	Stabilisation pathway to 7 W/m^2 (~800 ppm CO_2 between 2081 and 2100)
RCP4.5	Stabilisation pathway to 4.5 W/m^2 (~600 ppm CO_2 between 2081 and 2100)
RCP2.6 (also called RCP3PD)	Peak in radiative forcing at ~3 W/m^2 (peak of ~490 ppm CO_2 and then negative emission to ~450 ppm CO_2 between 2081 and 2100 leading to 2.6 W/m^2 by 2100)
RCP1.9 (also called the 1.5°C scenario)	Peak in radiative forcing at ~1.9 W/m^2 (~400 ppm CO_2 between 2081 and 2100; all the SSPs require negative global emissions to achieve this pathway)

providing a clear narrative and an outcome. This is because the SSPs do not include any mitigation measures, so a high emission global economy could have a lower concentration pathway by employing huge amounts of mitigation. If all the SSP and RCP combinations are examined, then some are extremely unlikely, while some are almost impossible, for example, SSP5 and RCP1.9. The Sixth Assessment report focuses on five main scenarios SSP1-1.9, SSP1-2.6, SSP2-4.5, SSP3-7.0, and SSP5-8.5.

Modelling Uncertainty

In the most recent IPCC Sixth Assessment report[2], the past and future emission scenarios were used in about 100 distinct independent GCMs. Each of these models has their own independent design and parameterisations of key processes. The independence of each model is important, as confidence may be derived from multiple runs on different models providing similar future climate predictions. In addition, the differences between the models can help us to learn about their individual limitations and advantages. Within the IPCC, due to political expediency, each model and its output is assumed to be equally valid. This is despite the fact that some are known to perform better than others when tested against the reality provided by the historical and paleoclimate records. Moreover, though we understand uncertainty within a single model, the notion of quantifying uncertainty from many models currently lacks any real theoretical background or basis[8]. The IPCC combines all the models used for each run and then presents the mean and the uncertainty between the models. In this way, it is clear that there are differences in the model output, but that in general, they agree and show very different futures based on which scenario we take. The uncertainties in the IPCC 2021 report are slightly higher than those in previous reports. This is because we achieve a greater understanding of the processes and improve our ability to quantify the uncertainty of our knowledge. So, although our confidence in the climate models has increased, so has the range of possible answers for any specific GHG forcing. One way to test the models and their uncertainty is to compare their predictions with the real-world outcome. The Coupled Model Intercomparison Project 3 took place just before the IPCC Third assessment report in 2001, and we can use those model predictions to compare with the following twenty years of real data. As we can see from Figure 37, the

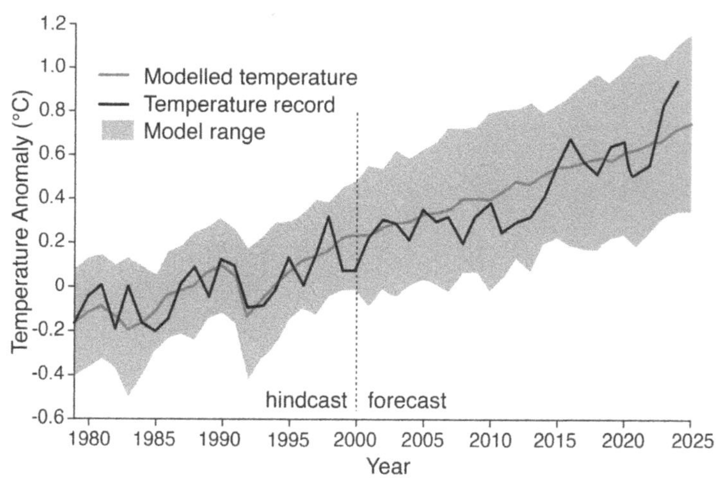

Figure 37. Climate Model Predictions Compared with Climate Data (1978 to 2024)

predicted warming of the world by the earlier climate models was very good, and we have now had over twenty years to improve the models and expand the number we use in our predictions.

Future Global Temperatures, Precipitation, Sea Ice, and Sea Level

Between 10 and 36 climate models have been run for each of the five SSP-RCPs for the IPCC 2021 report to produce scenarios of global temperature, precipitation, sea ice and sea level changes[2] that may occur by 2100. These climate models suggest that depending on our GHG emissions, the global surface temperature could, by 2100, rise between 1.5°C and 5.5°C compared with the pre-industrial period (1850–1900); you can refer to Table 3 for details. In all the scenarios except SSP1-1.9, a global temperature rise of over 1.5°C is reached between 2021–2041 with the best estimate being 2030. This is in line with observations as 2024 was 1.6°C warmer than pre-industrial times according to some temperature compilations and was above the 1.5°C limit in all global compilations. The models also show that the temperature rise will be unevenly distributed, with the largest rises in temperature observed on land.

Future sea-level rise is dependent on which SSP we follow and could be between 0.52 m and 1.02 m. This is similar but more extreme than the projection made by the previous IPCC report in 2007, which suggested a sea level rise of between 28 cm and 79 cm by 2100 (Figure 38).

Table 3. Temperature, Precipitation and Sea Level Projections[2]

Shared Socioeconomic Pathways	2081–2100 Global temperature changes relative to the pre-industrial period (°C)	2081–2100 Global sea level rise relative to 1990–2014 (m)	2081–2100 Global precipitation rise relative to 1990–2014 (%)
SSP5-8.5	4.7 (3.4 to 5.5)	0.73 (0.50 to 1.07)	8.2% (2.5 to 13.8%)
SSP3-7.0	3.9 (2.8 to 4.6)	0.65 (0.41 to 1.00)	5.5% (0.5 to 10.4%)
SSP2-4.5	2.9 (2.1 to 3.3)	0.55 (0.28 to 0.89)	4.7% (1.7 to 6.4%)
SSP1-2.6	1.9 (1.4 to 2.2)	0.41 (0.29 to 0.71)	3.2% (0.7 to 5.6%)
SSP1-1.9	1.5 (1.1 to 1.7)	0.35 (0.28 to 0.55)	2.7% (0.6 to 4.8%)

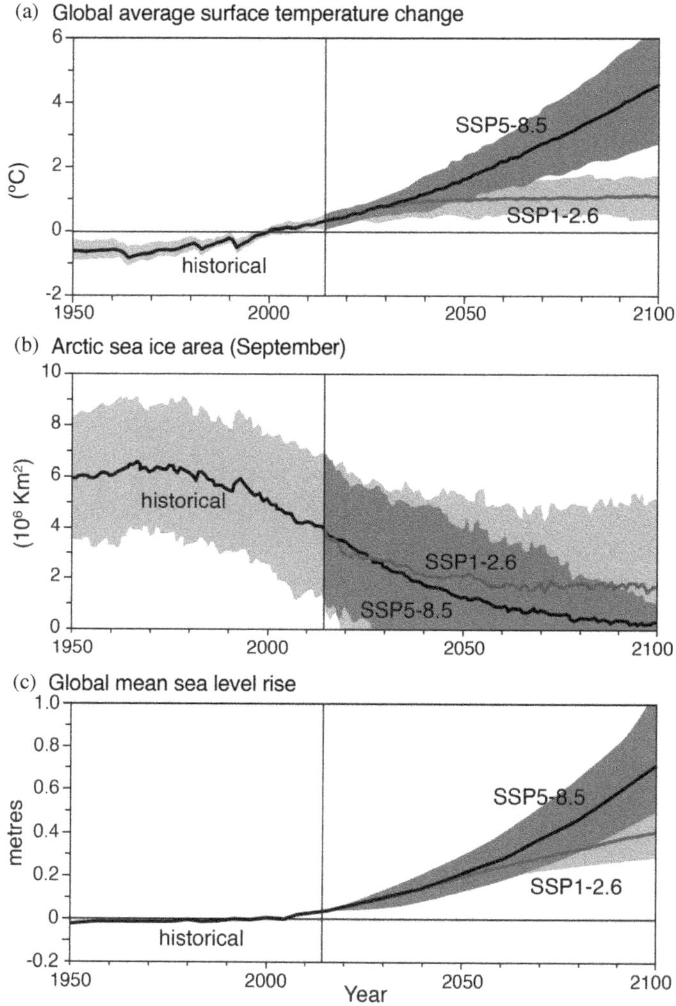

Figure 38. Global Temperatures, Arctic Sea Ice, and Sea Level in the 21st Century[2]

Both average land and ocean precipitation are very likely to increase under all five of the SSPs (see Table 3). The annual global average land precipitation by 2081–2100 relative to 1995–2014 will increase by 2.7% (with a range of 0.6–4.8%) in the low-emission scenario SSP1-1.9 and 8.2% (with a range of 2.5–13.8%) in the high-emission scenario

SSP5-8.5. Based on all the scenarios, the average land precipitation will increase approximately 1–3% per 1°C of global warming.

In the three worst-case SSP scenarios (SSP2-4.5, SSP3-7.0, and SSP5-8.5), the Arctic Ocean will become effectively ice-free (coverage below 1 million km^2) in September (the minimum ice month) by 2100.

Testing the Models

Different models give different results, so how can we trust any of them? This is a frequent response from many people not familiar with modelling, as there is a feeling that somehow science must be able to predict an exact future. However, in no walks of life can we expect this precision. For example, you would never expect to get a perfect prediction of which horse will win a race or which football team will emerge triumphant. The truth is that none of the climate models is right, because they only provide a range of potential futures. Our view of the future is strengthened by the use of more than one model, because each model has been developed by different groups of scientists around the world, using different assumptions, different computers, different programming languages; thus, they provide their own independent future predictions. What causes scientists to have confidence in the model results is that they all predict the same trend in global temperature and sea level for the next 80 years. One of the great strengths of the 2021 IPCC reports is that they used over 100 distinct models from 49 different modelling groups around the world compared to 40 models in 2013, 23 models in 2007 IPCC and 7 in 2001.

To dismiss the importance of climate change, many climate change deniers argue that the models are overly-sensitive to changes in GHGs. This is a classic 'it is not as bad as you think' argument. The advantage of having so many climate models is that scientists can also give an estimation of how confident they are in the model results and check how sensitive their models are compared with one another and the real-world data. One key test of climate models is the equilibrium climate sensitivity (ECS), whereby the model predicts how the global temperature change would be if pre-industrial carbon dioxide levels were doubled. These results have been very consistent over the past 50 years, and the 2021

IPCC report suggests the most up to date modelled range is between 2.50°C and 5.43°C (average 3.74°C), which is consistent with other measures.

Many climate change deniers argue that the current warming trend is due to natural variations. But scientists include all the natural variables, including ones that cool the climate in the climate models. Combining all our scientific knowledge of natural (solar, volcanic, aerosols, and ozone) and human-made (greenhouse gases and land-use changes) factors that are warming and cooling the climate, the result shows that 100% of the warming observed over the past 150 years is due to humans.

One of the uncertainties in the models is the role of the clouds and their interaction with radiation; this has been the case since the very first IPCC report in 1990. Clouds can both absorb and reflect radiation, thereby cooling the surface, as well as absorb and emit long-wave radiation, warming the surface. The competition between these effects depends on a number of factors: the height, thickness, and radiative properties of clouds. The radiative properties, formation and development of clouds depend on the distribution of atmospheric water vapour, water drops, ice particles, atmospheric aerosols, and cloud thickness. The physical basis of how clouds are represented or parametrised in the climate models has greatly improved through the inclusion of representations of cloud microphysical properties in the cloud water budget equations. Figure 34 above shows that even if the most extreme cooling value is applied for clouds, the warming factors due to GHGs are still three times larger.

Summary

Modelling future climate change is about understanding the fundamental physical processes of the climate system. Five new emission scenarios were produced for the 2021 IPCC Science report, using a much wider set of inputs to the socioeconomic models including population, land use, energy intensity, energy use, and regional differentiated development. One of these emissions pathways (SSP1-1.9) was developed to indicate to policy makers how they could achieve the aspired target of just 1.5°C warming set at the Paris 2015 climate change conference. Over 100 climate models were used in developing the IPCC scenarios, providing a

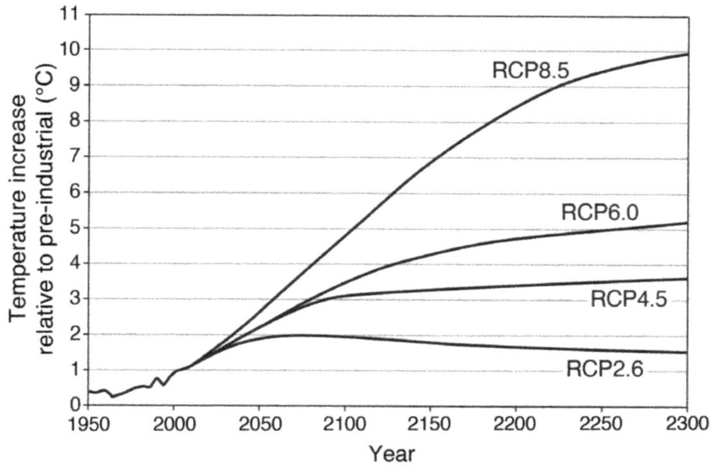

Figure 39. Global Surface Temperatures (1950–2300)

huge 'weight of evidence'. Using the three main realistic carbon emissions pathways over the next 80 years, the climate models suggest the global mean surface temperature could rise by between 2.1°C and 5.5°C by 2100. However, it must remember that global temperatures will not stop changing once we get to year 2100. Figure 39 shows how temperatures could continue to rise way beyond the levels of this century, depending on the chosen emission pathway. Using the three main realistic carbon emissions pathways, the models also predict an increase in global mean sea level of between 0.50 m and 1.3 m by 2100 compared with pre-industrial times.

Chapter 7

Climate Change Impacts

This chapter assesses the potential impacts of climate change and how they alter in scale and intensify with increasing warming. The IPCC 'Impacts, Adaptation and Vulnerability' report[1] looks at potential impacts on a regional level as well as by different sectors, such as freshwater resources, ecosystems, coastal and ocean systems, food security, and human health. It is also necessary to estimate the extent and magnitude of climate change impacts at national and local levels. There are many excellent national reports and tools, such as the Singapore's National Climate Change Study[2], the Korean Climate Change Assessment Report[3], The People's Republic of China's National Communication on Climate Change[4], the US National Climate Assessment[5] and the UK Climate Impacts Programme[6]. All of these reports have interactive tools to understand the potential effects of climate change within their own countries. In this chapter, the potential impacts are broken down into sectors: extreme heat and droughts, storms and floods, agriculture, ocean acidification, biodiversity, and human health.

What is Dangerous Climate Change?

One of the most important questions for policy makers is "what is dangerous climate change?" It is a vital question because if we are committed to cut global GHG emissions, we need a realistic target concerning the degree of climate change with which we can cope. In February 2005, the British government convened an international science meeting at Exeter,

UK, to discuss this very topic. This was a very political science meeting, as the UK government was looking for a recommendation to take to the G8 meeting in Gleneagles. At that time, Britain held both the chair of the G8 and presidency of the EU, and the then Prime Minister, Tony Blair wished to push forward internationally his joint agenda of climate change reduction and poverty alleviation in Africa. The meeting and a lot of supporting research at the time suggested a limit of 2°C above pre-industrial average temperature: below this threshold, there seemed to be both winners and losers due to regional climate change, but above this temperature, everyone seemed to lose. It has now been shown that due to the impacts of extreme weather events, there are in fact no regions that benefit from a 2°C warming[7]. At the Paris climate change negotiation meeting in 2015, the alliance of Small Island Nations and some key developing countries reiterated that even a small amount of warming would be devastating for their countries (see Chapter 10 for more details). The Paris agreement set 2°C as the key target but added the aspiration of 1.5°C. Subsequently, the IPCC Special Report on 1.5°C global warming published in 2018 supported this lower target by demonstrating that there is a significant increase in regional and national climate change impacts between an increase of a 1.5°C and a 2.0°C world.

Extreme Events and Society's Coping Range

The single biggest problem with climate change is our inability to predict the future. Humanity can live, survive, and even flourish in extreme climates from the Arctic to the Gobi Desert, but problems arise when the predictable extremes of local climate are exceeded. For example, heat waves, storms, droughts, and floods in one region may be considered fairly normal weather in another. This is because each society has a coping range, defined as a range of weather with which it can deal with. Figure 40 shows the theoretical effect of combining the societal coping range with climate change. In our present climate, the coping range encompasses nearly all the variation in weather with maybe only one or two extreme events. These could be 1-in-200-year events that surpass the coping ability of that society. As the climate moves gently to its new average, more

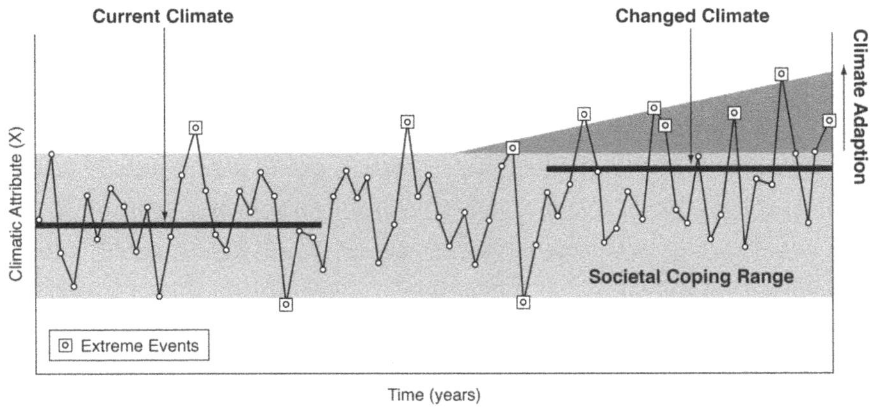

Figure 40. Climate Change, Societal Coping Range, and Extreme Events

extreme events will occur if the coping range stays the same. Hence a 1-in-200-year event may become a 1-in-50-year event. The good news is that the societal coping range is flexible and can change to cope with the shifting baseline and the more frequent extreme events — as long as there is strong climate science to provide clear guidance on the sort of changes that are going to occur. The speed with which the societal coping range can vary depends on what aspect of society is being affected. For example, adaptation of the individual's behaviour can be extremely quick, while building major infrastructure can take decades to complete. In fact, one of the biggest challenges of climate change is to build flexible and resilient societies that are able to adapt to a changing future.

Extreme Heat, Drought, and Wildfires

As global temperatures increase, heat waves increase. As precipitation becomes more variable and concentrated into more intense rainfall events, dry periods get longer and droughts increase. The combination of extreme heat and drought causes more wildfires.

Heat waves are often referred to as the 'silent killer'. They disproportionally affect the elderly, and it is the sustained night-time temperatures that kill because while asleep, older people are less able to regulate their body temperature. The Lancet Countdown 2025 report has tracked global heat-related mortality[8]. Over the last five-year average, 84% (16 out of 19) of the life-threatening extreme heatwaves would have not occurred without climate change. The most vulnerable age groups, those under 1 year old and those over 65 years old were exposed to a record number of heatwave days in 2024, with 389% and 304% more heatwave days per person, respectively, compared to 30 years ago. Over the last decade, the higher temperatures have led to a 63% increase in heat-related deaths since the 1990s, reaching an estimated 546,000 deaths per year. Heat exposure also endangers workers' health, particularly those working outdoors: for example, in construction or agriculture. It can reduce productivity, income and livelihoods, as well as negatively impacting health. In 2024, a record-high of over 640 billion potential work hours were affected, a 98% increase compared to 30 years ago.

Heat waves and droughts, however, are relative terms, as it depends on where they occur, and whether a region already has adaptations in place. The 2003 heat wave in Europe killed an estimated 70,000 people[9] with the hardest hit in France with 14,800 deaths in the first three weeks of August and deaths in Paris increased by 140%. After the 2003 heat wave, it was realised that many of these deaths were due to the very weak public health response. As a result, there were sweeping policy changes in many countries, including better heat wave prediction and emergency preparations, improved building design and air conditioning for hospitals and retirement homes, increased training for health care professionals, an emphasis on responsible media coverage with health recommendations, as well as planned regular visits to the most vulnerable members of the population. These new policies throughout Europe have prevented a significant number of deaths during subsequent heat waves, such as 2018. One of the reasons it is so difficult to understand the impacts of climate change is because people and societies can adapt to new conditions very quickly. As climate change increases, it is clear that the temperature of the 2003 heat wave could be the average summer temperature in 2050 and heat waves above this new baseline may still occur. Adaptation to heat waves, however, takes planning, resources and money. While this has been possible in much of Europe, there are many areas of the world where such preparation is not happening due to poverty and lack of good governance.

Droughts are also a major killer that should be considered. A drought happens when an area undergoes a prolonged period without sufficient water supply, whether surface or underground water. A drought can last for months or years, and is usually caused when an area receives consistently below average rainfall. Droughts have a substantial impact on the local ecosystem and agriculture, including drops in crop growth and yield, as well as the loss of livestock[8]. Although droughts can persist for several years, even a short, intense drought can cause significant damage and harm to the local economy. Prolonged drought has caused famine, mass migrations, and humanitarian crises. From a disease point of view, droughts are much worse than floods because of a lack of fresh drinking water, and stagnant pools of water can bring disease. In 2024, the percentage of global land area affected by at least one month of extreme drought reached a record-breaking 48%. This is nearly a 300% increase over the last 70 years. Globally, over 25% of the land was exposed to extreme drought, compared to 10% in 1900s, with the Amazonian region, south, north and east Africa, as well as the Horn of Africa in particular, disproportionately affected (Figure 41).

There has been an increase in wildfire risk in 114 out of 196 countries over the last 10 years. The most significant increases have occurred in the Southern Hemisphere, Africa, and Brazil. Over the last five years, exposure to very high wildfire risk has increased by 6.6% on average globally, or six days per person. In 2024, there were 154,000 deaths

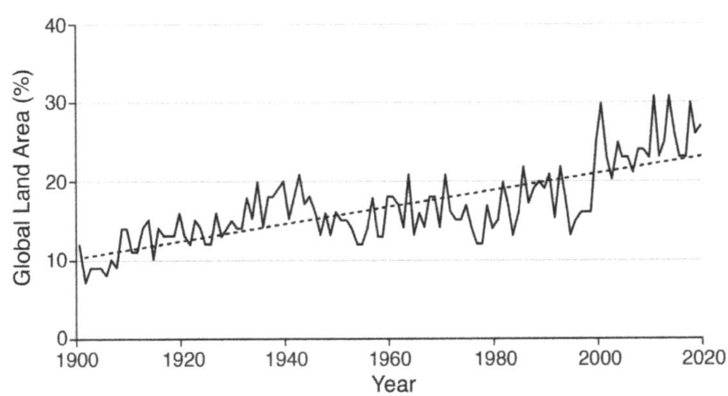

Figure 41. Global Increase in Land Area Affected by Droughts[10]

from wildfire-derived $PM_{2.5}$ particulate air pollution, up by 36% from 10 years ago[8].

Of most note was the largest wildfire in 2003 that occurred in the Siberian Taiga that burned approximately 86 million acres of land across Siberia, the Russian Far East, and parts of China and Mongolia. In 2023, the South American wildfires affected Brazil, Bolivia, Chile, Colombia, Ecuador, Paraguay, Peru, and Venezuela, burning over 330,00 square miles. Canada and USA record-breaking wildfires happened in 2015, 2020, 2023 and 2025. In 2019–20, the Australian bushfires season became known as the 'Black Summer'. Throughout the summer, hundreds of fires burned, mainly in the southeast of the country, due to record heat and a prolonged drought. The fires burned an estimated 154,000 square miles, destroying nearly 10,000 buildings, and killing at least 450 people and over one billion animals. But even this bushfires season was eclipsed by the 2023/24 Australian wildfires which burnt over 555,000 square miles. These wildfires have pushed endangered species to extinction. With increasing extreme heat and droughts, wildfire risk will continue to grow across the world.

Storms and Floods

Storms and floods are major natural hazards. Over the past two decades, they have been responsible for three-quarters of the global insured losses, and over half the fatalities and economic losses from natural catastrophes. It is, therefore, essential that we know what is likely to happen in the future. There is evidence that the temperate regions, particularly in the Northern Hemisphere, have become stormier over the past 50 years. Flood events have shown an upward trend since 2005; in the last decade, a record 64% of global land area saw an increase in extreme precipitation events from 50 years ago. This has not been driven by increases in any specific areas, but by an increase in a wide range of events across the globe. The climate models suggest that the proportion of heavy rainfall has and will continue to increase, as will the year-to-year variability. This will increase the frequency and magnitude of flooding events.

Two-fifths of the world's population lives under the monsoon belt, which brings life-giving rains. Monsoons are driven by the temperature contrast between continents and oceans. Moisture-laden surface air blows from the Indian Ocean to the Asian continent, and from the Atlantic Ocean into West Africa during Northern Hemisphere summers, when the land masses become much warmer than the adjacent ocean. In winter, the continents become cooler than the adjacent oceans while high pressure develops at the surface, causing surface winds to blow towards the ocean. Climate models indicate an increase in the strength of the summer monsoons as a result of global warming over the next 100 years. There are three reasons why this should occur: (1) global warming will cause temperatures on continents to rise even higher than those of the ocean in summer and this is the primary driving force of the monsoon system; (2) decreased snow cover in Tibet, which is to be expected in a warmer world, will increase this temperature difference between land and sea, increasing the strength of the Asian summer monsoon; (3) a warmer climate means the air can hold more water vapour, so the monsoon winds will be able to carry more moisture. For the Asian summer monsoon, this could mean an increase of 10–20% in average rainfall, with an interannual variability of 25–100% and a dramatic increase in the number of days with heavy rain. The most concerning finding of the models is the predicted increase in rain variability between years, which could double, making it very difficult to predict how much rainfall will occur each year — essential knowledge for farmers.

One of the more contentious areas of climate change science is the study and predictions of future tropical cyclones, better known as typhoons or hurricanes. There is clear evidence that the number and intensity of hurricanes have increased over the past four decades in the North Atlantic Ocean, southern Indian Ocean, and Pacific Ocean. This is because the number and intensity of hurricanes are directly linked to the sea-surface temperature. As hurricanes can only start to form if the sea-surface temperature is above 26°C, it would seem logical that in a warmer world, there would be more hurricanes[11]. Yet the actual formation of hurricanes is much rarer than the opportunities for them to occur. Only 10% of centres of falling pressure over the tropical oceans give rise to fully-fledged hurricanes. Other considerations, such as wind shear to start the rising air spinning, must be included to understand the genesis of tropical storms. In a year of high incident, perhaps a maximum of 50 tropical storms will develop to hurricane levels. Predicting the level of a disaster is difficult, as the number of hurricanes is not the key; it is whether they make landfall, and how intense and prolonged they are once they have hit land.

When hurricanes hit in developed countries, the major effect is usually economic loss, while in developing countries, it is the loss of life. For example, Hurricane Katrina, which hit New Orleans in 2005, resulted in 1,800 deaths and caused over $150 billion in damage. By contrast, in 2013, Typhoon Haiyan devastated large portions of Southeast Asia, particularly the Philippines, affecting 11 million people, causing over 6,300 deaths with another 1000 people missing, but only resulted in a reported $2.2 billion in damages.

One of the most important parts of the global climate system is the periodic switch of the direction and intensity of ocean currents and winds in the Pacific, known as ENSO (El Niño–Southern Oscillation). This phenomenon typically occurs every three to seven years and may last from several months to more than a year. ENSO is an oscillation between three climates: 'normal' conditions, La Niña (a cooler opposite of El Niño), and El Niño. El Niño conditions have been linked to changes in the monsoon, storm patterns, and occurrence of droughts all over the world. El Niño conditions cause droughts in the southern USA, East Africa, northern India, north-east Brazil, and Australia. While in Indonesia, forest wildfires are more common because of the very dry conditions. In California, parts of South America, Sri Lanka, and East Central Africa, El Niño is linked to torrential rains and extreme floods.

The state of the ENSO has also been linked to the position and occurrence of hurricanes in the Atlantic. There is also considerable debate over whether ENSO has been affected by climate change. Reconstruction of past climate in the western Pacific using coral reefs shows sea-surface temperature variations dating back 150 years, well beyond our historical records. The sea-surface temperature shows the shifts in ocean current, which accompany shifts in the ENSO and reveal that there have been two major changes in the frequency and intensity of El Niño events. First was a shift at the beginning of the 20th century from a 10–15-year cycle to a 3–7-year cycle. The second was a sharp threshold in 1976 when a marked shift to more intense and frequent El Niño events occurred with a cycle of between 2 and 4 years. Climate models all agree that ENSO will continue in the future, and in the higher emission scenarios in the second half of this century, ENSO variability will become much more extreme, creating more rainfall and drought events as well as influencing the number and intensity of tropical storms in unpredictable ways.

Coasts

As we have seen, the IPCC reports that sea level could rise by between 50 cm and 130 cm by 2100 compared with pre-industrial times[12]. This prediction is of major concern to all who live in coastal areas, as rising sea levels will reduce the effectiveness of coastal defenses against storms and floods, as well as increase the instability of cliffs and beaches. In the developed world, the response to this danger is to add another few feet to the height of the sea walls around property on the coast, the abandoning of some poorer quality agricultural land to the sea (as it is no longer worth the expense of protecting it), and the enhancement of legal protection for coastal wetlands, being nature's best defense against the sea. Globally, there are some nations based on small islands and river deltas that face a much more urgent situation (see Figure 42).

For small island nations, such as the Maldives in the Indian Ocean and the Marshall Islands in the Pacific, a 1 m rise in sea level would flood up

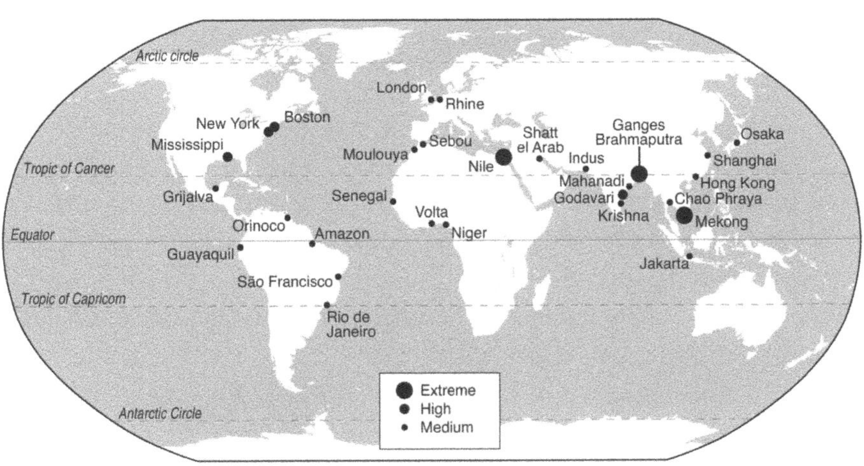

Figure 42. Areas Most at Risk from Sea-Level Rise

to 75% of the dry land, making the islands uninhabitable[12]. Interestingly, it is also these countries, which rely mainly on tourism, that have some of the highest fossil-fuel emissions per head of population. Other major concentrations of population at risk are those who live alongside river deltas, as in, for example, Bangladesh, Egypt, Nigeria, and Thailand. A World Bank report concluded that human activities on the deltas, such as dams and fresh-water extraction, were causing these areas to sink much faster than any predicted rise in sea level, increasing their vulnerability to storms and floods.

In the case of Bangladesh, over three-quarters of the country is within the deltaic region formed by the confluence of the Ganges, Brahmaputra, and Meghna rivers. Over half the country lies less than 5m above sea level, so flooding is a common occurrence. During the summer monsoon, a quarter of the country is flooded. Yet these floods, like those of the Nile, bring with them life as well as destruction. The water irrigates and the silt fertilises the land. The fertile Bengal delta supports one of the world's most dense populations, over 110 million people in 140,000 square kilometres (km^2). Every year, the Bengal delta is supposed to receive over one billion tonnes of sediment and 1,000 cubic kilometres (km^3) of fresh water. This sediment load balances the erosion of the delta both by natural processes and human activity. However, the Ganges, Brahmaputra, and Meghna rivers have been dammed for irrigation and power generation, preventing the movement of silt downriver. The reduced sediment input is causing the delta to subside. Exacerbating this is the rapid extraction of fresh water.

Since the 1980s, 100,000 tube wells and 20,000 deep wells have been sunk, increasing the fresh-water extraction six-fold. These wells are essential to improving the quality of life for people in this region, but have produced a subsidence rate of up to 2.5 cm per year, one of the highest rates in the world. Using estimates of subsidence rate and global warming sea-level rise, the World Bank has estimated that by the end of the 21st century, the relative sea level in Bangladesh could rise by as much as 1.8 m. In a worst-case scenario, they estimated that this would result in a loss of up to 16% of land, supporting 13% of the population, and producing 12% of the current GDP. Unfortunately, this scenario does not take any account of the devastation of the mangrove forest and the associated fisheries. Moreover, increased landward intrusions of salt water would further damage water quality and agriculture.

Many major cities around the world are vulnerable to flooding because they were built close to rivers or the coast in order to facilitate trade via the oceans. Examples of current cities most at risk include in Asia: Dhaka (20.3 million people today), Shanghai (17.5 million), Guangzhou (13 million), Shenzhen (12.5 million), Jakarta (10.8 million), Bangkok (10.5 million), Hong Kong (8.4 million), Ho Chi Minh City (8.3 million) and Osaka (5.2 million); in North America: New York (18.8 million), Boston (4.9), Miami (2.7 million) and New Orleans (0.4 million); in South America: Guayaquil, (2.9 million) and Rio de Janeiro (1.8 million); in Africa: Abidjan (3.7 million) and Alexandria (3.0 million); and in Europe: London (8.9 million) and Hague (2.5 million). The annual Lancet Countdown Report estimated that without intervention, between 145 million and 565 million people could be affected and displaced as a result of future sea-level rise.

Agriculture

One of the major worries concerning climate change is the effect it will have on agriculture, both globally and regionally[13]. The main question is whether the world can feed itself with an extra two billion people on the planet by 2050 and a rapidly changing climate. Figure 43 shows the drop in cereal grain yields that have already occurred. Modeling suggests that in higher latitudes, agricultural productivity may increase due to the longer growing season and reduced frost damage — but some of this will be offset by more frequent crop damage due to extreme weather events. Agriculture production will, however, reduce significantly in the Tropics and sub-Tropics due to much hotter temperatures and more variable rainfall.

Higher temperatures and humidity will also be a challenge for many societies that rely heavily on subsistence agriculture, as higher air temperatures and humidity make working outside more challenging and increase the likelihood of hyperthermia[8]. This will impact on the health of anyone who has to work outside regularly, including construction and farm workers. In 2024, a record 640 billion potential work hours were lost.

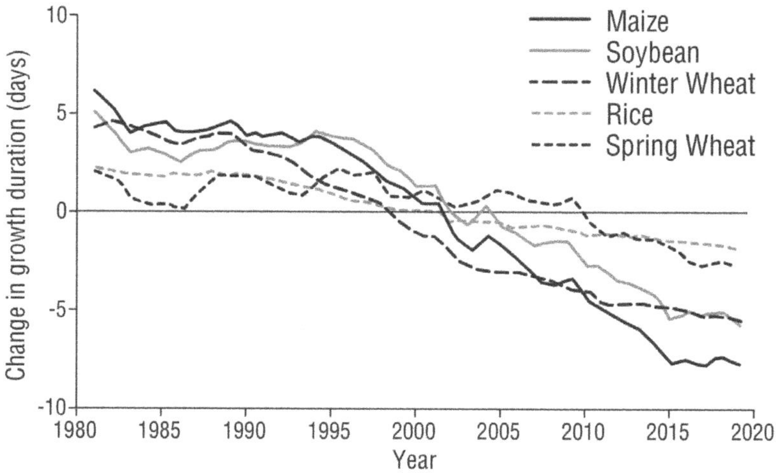

Figure 43. Changes in Cereal Grain Yields between 2080 and 2020

This is a 98% increase compared to 30 years ago or 454 billion hours greater than in 2000. Seven countries (Cambodia, India, China, Indonesia, Nigeria, Brazil and the USA) together represent approximately 60% of the global hours lost, with India experiencing by far the greatest loss. In the first six countries, the impact on work hours lost fell mainly on agricultural workers.

Estimating the overall impact of climate change on agriculture is difficult as agricultural production has very little to do with feeding the world's population and much more to do with trade and economics. This is why the European Union has stockpiles of food, while many underdeveloped countries export cash crops (such as sugar, cocoa, coffee, tea, and rubber) but cannot adequately feed their own populations. A classic example is the West African state of Benin, where cotton farmers can obtain cotton yields four to eight times per hectare greater than their US competitors in Texas. However, because the USA subsidises its farmers, this means that US cotton is cheaper than that coming from Benin. Currently, US cotton farmers receive over $4 billion in subsidies, almost twice the total GDP of Benin. In 2002, Brazil filed a case with the World Trade Organization (WTO) against the USA for unfair subsidies and distortion of trade. They won their case in 2005; however, over 20 years later, the

USA is still discussing what changes should be made to their farming subsidies. So, even if climate change makes Texan cotton yields even lower, it does not change the biased market forces still illegally in operation.

Markets can reinforce the difference between agricultural impacts in developed and developing countries. Variations in supply and demand could mean that agricultural exporters may gain in monetary terms even though the supplies fall, because when a product becomes scarce, the price rises. The other completely unknown factor is the extent to which a country's agriculture can be adapted. For example, in climate change models, it is assumed that production levels in developing countries will fall to a greater degree than those in developed countries because their estimated capability to adapt is less than that of developed countries. This is just another assumption that has no analogue in the past, and as these effects on agriculture will occur over the next century, many developing countries may catch up with the developed world in terms of adaptability.

Ocean Acidification

Direct measurements of the ocean's chemistry have shown that the pH of the oceans is getting lower; that is, they are getting more acidic (see Figure 44)[14]. This is because carbon dioxide in the atmosphere dissolves

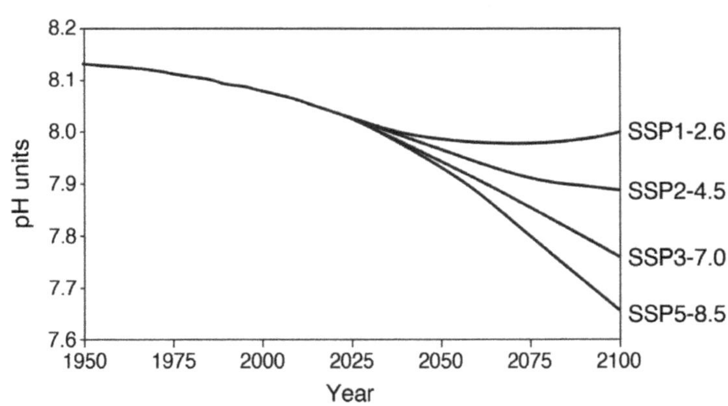

Figure 44. Ocean Acidification[14]

in the water of the surface ocean. This process is controlled by two main factors: the amount of carbon dioxide in the atmosphere and the temperature of the ocean. The oceans have already absorbed about a third of the carbon dioxide resulting from human activities, which has led to a steady decrease in ocean pH levels. With increasing atmospheric carbon dioxide in the future, the amount of dissolved carbon dioxide in the ocean will continue to increase. Some marine organisms, such as corals, foraminifera, coccoliths, and shellfish, have shells composed of calcium carbonate, which dissolves in acid. Laboratory and field experiments show that with high level of carbon dioxide, the more acidic waters cause some marine species to develop misshapen shells and have lower growth rates, although the effect varies among species. Acidification also alters the cycling of nutrients and many other elements and compounds in the ocean, leading to a likely shift in the competitive advantage among species, and also have impacts on marine ecosystems and the food web. This is a major worry as fishing is still a major source of food, with about 95 million tonnes of fish caught by commercial fishing and another 50 million tonnes produced by fish farms per year.

Biodiversity

The current loss of biodiversity around the world is due to human activity, including deforestation, agriculture, urbanisation, and mineral exploitation. Extinction rates are currently 100–1,000 times higher than the background natural rate and climate change will exacerbate this decline[15]. The IPCC impact reports[1] lists the following species as those most at risk from climate change: the mountain gorilla in Africa; amphibians that live in the cloud forests of the neotropics; the spectacled bear of the Andes; forest birds of Tanzania; the 'resplendent quetzal' bird in Central America; the Bengal tiger, and other species found only in the Sundarban wetlands; rainfall-sensitive plants found only in the Cape Floral Kingdom of South Africa; as well as polar bears and penguins near the poles. The primary reason for the threat to these species is that they are unable to migrate in response to climate change because of their particular geographical location or the encroachment of human activity, especially farming and urbanisation. An example of the former is the cloud forests of the neotropics: as climate changes, this particular climatic zone will migrate up the mountainside — until the point where there is no more mountain left.

One example of an ecosystem under threat is the coral reefs. Coral reefs are a valuable economic resource for fisheries, recreation, tourism, and coastal protection. Some estimate that the global cost of losing the coral reefs runs into hundreds of billions of dollars each year. In addition, reefs are one of the largest global stores of marine biodiversity. The last few years have seen unprecedented declines in the health of coral reefs. An estimated 50% of the hard corals on Australia's Great Barrier Reef have been lost in the past few years from bleaching events, which are caused by extreme water temperatures. The Great Barrier Reef is the world's largest coral reef system, composed of over 2,900 individual reefs and 900 islands stretching for over 1,400 miles. In other regions, as much as 70% of the coral has died in a single season. There has also been an upsurge in the variety, incidence, and virulence of coral disease in recent years, with major die-offs in Florida and much of the Caribbean region. In addition, increasing atmospheric carbon dioxide concentrations could decrease the calcification rates of the reef-building corals, resulting in weaker skeletons, reduced growth rates, and increased vulnerability to erosion. Model results suggest these effects would be most severe at the current margins of coral reef distribution.

On a more theoretical note, the biologist, Chris Thomas, and his colleagues published a study in *Nature* in 2004 investigating the possible increase in the extinction rates over the next 50 years in key regions such as Mexico, Amazonia, and Australia. Their theoretical models suggest that a 2°C warming by 2050 could condemn one-quarter of all species they studied to extinction[16]. This study has been criticised as there are many assumptions in their models which may or may not be true; for example, they assume we know the full climatic range in which each species can persist and the precise relationship between shrinking habitat and extinction rates. Hence, these results should be seen only as the likely direction of extinction rates, not necessarily the exact magnitude. However, this and many other scientific studies demonstrate the huge threat to regional and global biodiversity as well as illustrate the sensitivity of biological systems to the amount and rate of warming that will occur in the future. The crisis in biodiversity opens up another theoretical consideration which is how classical economics deal with infinite loss — this is a loss of a species, or ecosystem, or a culture which no amount of money can ever bring back. Economics has no way of dealing with the infinite cost of these terminal losses — so it ignores it.

Human Health

The potential health impacts of climate change are immense, and managing those impacts will be an enormous challenge. Climate change will increase deaths from heat waves, droughts, wildfires, storms, and floods[9]. Higher temperatures and variable precipitation threaten food production. This is due to reduced productivity because of the increased risks to people working outside regularly, including construction and farm workers. It has been suggested that the overall death rate may drop in some countries since many elderly people die from cold weather so warmer winters would reduce this cause of death. This claim is complicated because there are many factors at work. It is true that deaths from cold conditions are higher than those from hot conditions in most temperate counties, but improved housing, health care, higher incomes, and greater awareness of the cold risks have been responsible for the reduction of excess winter deaths in many of these countries, including in China and the UK. Hence, in many societies, adaptation to cold climate and improved protection for the most vulnerable members of society means that warmer winters will have little or no effect in reducing the death rate. In comparison, summer heat-related deaths have risen all around the world.

The 2009 University College London's report in the journal *Lancet*, 'Managing the Health Effects of Climate Change', identified the two major areas that could affect the health of billions of people: water and food[9]. The most important threat to human health is the lack of access to fresh drinking water. At present, there are an estimated 2.2 billion people who lack access to safely managed drinking water services and basic sanitation. Not only does the lack of water cause major health problems from dehydration, a large number of diseases and parasites are also present in dirty water. The rising worldwide human population, particularly those concentrated in urban areas, is putting a great strain on water resources.

The impacts of climate change — including changes in temperature, precipitation, and sea levels — are expected to have varying consequences for the availability of fresh water around the world. For example, changes in river run-off will affect the yields of rivers and reservoirs, which in turn, reduce the recharging of groundwater supplies. An increase in the rate of evaporation will also affect water supplies and contribute to the salination of irrigated agricultural lands. Rising sea levels may result in saline intrusion in coastal aquifers. Currently, approximately 2 billion people, one-quarter of the world's population, live in countries that are water-stressed. It has been suggested that if nothing is done to mitigate climate change, then up to 50% of the world population could live in countries experiencing water stress by 2050. Of these people, 80% *are* living in developing countries.

Climate change is likely to have the greatest impact in countries with a high ratio of relative use to available supply. Regions with abundant water supplies will get more than they want with increased flooding. As suggested above, computer models predict much heavier rains and thus major flood problems for Asia, Africa and Europe, whilst, paradoxically, countries that currently have little water (e.g. those relying on desalination) may be relatively unaffected. It will be countries in between, which have no history or infrastructure for dealing with water shortages, that will be the most affected. In central Asia, North Africa, and southern Africa, there will be even less rainfall and water quality will become increasingly degraded through higher temperatures and pollutant run-off. Add to this the predicted increased year-to-year variability in rainfall, and droughts will become more common. Hence, it is those countries that have been identified as most at risk which need to start planning now to conserve their water supplies and/or deal with the increased risks of flooding; because it is the lack of infrastructure to deal with drought and floods rather than the lack or abundance of water which causes the threat to human health.

Food security rests on three main pillars: 1) food availability — production volume, 2) access — food affordability, and 3) stability — constant supply of food. According to UN World Food Programme, we currently produce enough food to feed 10 billion people easily, enough to cover the predicted increase in population this century. Yet, there are approximately 750 million people on the brink of starvation today, up by approximately 152 million than in 2019. This is because they simply do not have enough money to buy food. Climate change threatens food availability and stability as it affects the production of food and other agricultural goods. Extreme weather events must also be considered. With an increasingly globalised economy, very few countries are self-sufficient in basic food and hence, food imports are very important. The cost of basic food items can be strongly influenced by global demand, national agricultural subsidies and export bans, as well as natural disasters, but the biggest influence is food speculation in the global markets. In 2008–9, there was a 60% rise in the price of food while in 2011–12, there was a 40% jump in price. Both events were linked to underinvestment in agriculture, increasing biofuel demand, shifting dietary patterns, and food speculation. Food prices have risen markedly since 2022 and have remained high due in part to the war in Ukraine. The inability of many people to afford basic food, leading to malnutrition and starvation, can be linked directly to the speculation on food prices in the global markets and to distant foreign wars.

Climate change will worsen a wide range of health burdens, in particularly cardiovascular, respiratory and renal conditions[17]. Climate change is linked to a rising burden and/or prevalence of non-communicable neurological diseases, including stroke, multiple sclerosis and Alzheimer's disease. Hospital admissions and mortality risk for many mental health disorders also increase with increased temperature and humidity, as well as diurnal temperature fluctuation. For example, higher diurnal temperature ranges in Taiwan are associated with an increased hospital admission risk for schizophrenia. In California (USA), 6°C increase in temperature within a day was associated with an approximately 6.5% increase in hospital visits for any mental health disorder, self-harm, intentional injury of another person, or homicide. In Bern (Switzerland), time-series data of nearly 90,000 hospitalisations for psychiatric disorders between 1973 and 2017 showed a 4% increase for every 10°C increase in daily temperature.

Another threat to human health is the transmission of infectious diseases, which is directly affected by climatic factors[8]. Climate change will particularly influence vector-borne diseases, that is, diseases that are carried by another organism, for example malaria, which is carried by mosquitoes. Infective agents and their vector organisms are sensitive to factors such as temperature, surface-water temperature, humidity, wind, soil moisture, and changes in forest distribution. It is, therefore, projected that climate change and altered weather patterns would affect the range (both altitude and latitude), intensity, and seasonality of many vector-borne and other infectious diseases. For example, there is a strong correlation between increased sea-surface temperature and sea level, and the severity of the cholera epidemics in Bangladesh. With predicted future climate change and consequent rise in Bangladesh's relative sea level, cholera epidemics could become more common.

In general, increased warmth and moisture caused by climate change will enhance transmission of diseases. While the potential transmission of many of these diseases increases in response to climate change, we should remember that our capacity to control the diseases will also change. New or improved vaccination can be expected; some vector species can be constrained by use of pesticides. Nevertheless, there are uncertainties and risks here too: for example, long-term pesticide use encourages the breeding of resistant strains, while killing many natural predators of pests.

We now have good evidence for the increase of infectious diseases due to climate change. The average climate-defined transmission potential of dengue by *Aedes albopictus* and *A. aegypti* increased by 48.5% and 17.0%, between 1951–60 and 2015–24, contributing to over 14 million dengue cases reported globally (Figure 45)[8]. The sand fly transmitted *leishmaniasis* is endemic in 99 countries or territories. It disproportionately

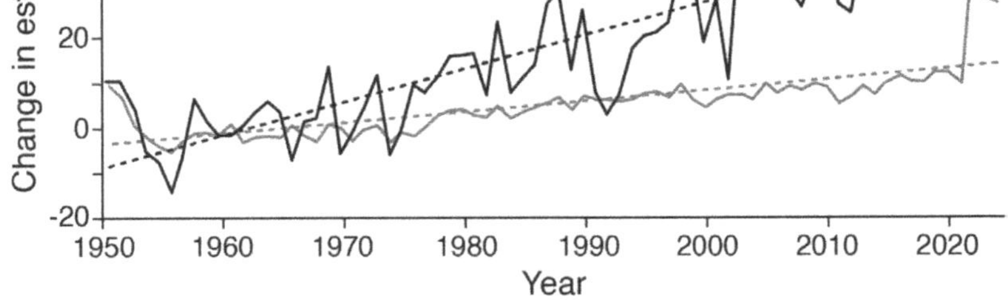

Figure 45. Changes in the Reproduction Rate (R0) of Dengue Fever[8]

affects the poorest people, with an estimated 700,000 to 1 million new cases every year causing 20,000–40,000 deaths over the last decade. Over the last 75 years, there has been an increase occurrence of 50% in prevalent countries. Also, an additional 364 million people were at risk of *Rhipicephalus sanguineus* and *Hyalomma* ticks-transmitted disease (including Rocky Mountain Spotted Fever and Crimean-Congo Hemorrhagic Fever) compared to the 1950s.

The most important vector-borne disease is malaria, with currently 500 million infected people worldwide. *Plasmodium vivax*, which is carried by the *Anopheles* mosquito, is the organism that causes malaria. The main climate factors that have a bearing on the malarial transmission potential of the mosquito population are temperature and precipitation. Assessments of the potential impact of global climate change on the incidence of malaria suggest a widespread increase of risk because of the expansion of the areas suitable for malaria transmission. Already in the past 5 years, the area suitable for malaria transmission in highland areas was 39% higher in Africa and 150% higher in East Asia compared to the 1950s. Mathematical models mapping out the temperature zones suitable for mosquitoes suggest that by the 2080s, the potential exposure of people could increase by 2–4% (260–320 million people). The predicted increase is most pronounced at the borders of endemic malarial areas and at higher altitudes within malarial areas. The changes in malaria risk must be interpreted on the basis of local environmental conditions, the effects of socioeconomic development, and malaria control programmes or capabilities. Climate change will also provide excellent conditions for *Anopheles* mosquitoes to breed in temperate areas such as Jeju Island (South Korea), Northern China, Nepal, northern India, Vietnam, England, continental Europe, and northern USA.

It should, however, be noted that the occurrence of many tropical diseases is also related to development. As recently as the 1940s, malaria was endemic in Finland, Poland, Russia, and 36 states in the USA including Washington, Oregon, Idaho, Montana, North Dakota, New York, Pennsylvania, and New Jersey. Although climate change has the potential to increase the range of many of these tropical diseases, the experiences of Europe and the USA suggest that combating malaria is strongly linked to development and resources: development to ensure efficient monitoring of the disease and resources to secure a strong effort to eradicate the mosquitoes and their breeding grounds.

Summary

The impacts of climate change will increase significantly as the temperature of the planet rises. Climate change will affect the return period and

Table 4. Potential Impacts of Climate Change

POTENTIAL IMPACTS OF CLIMATE CHANGE

°C above pre-industrial levels	Impacts
1.5°C	• Major effects on warm water coral reef ecosystem. • Significant impacts on vulnerable ecosystems and species (polar regions, wetlands, and cloud forests). • Increase in coastal and river flooding. • Increase in extreme weather events. • Increase in the spread of tropical infectious disease. • Increase in heat-related morbidity and mortality.
2°- 3°C	• The Maldives, the Marshall Islands, Tuvalu, and many other small island nations have been abandoned. • Major loss of warm water coral reef ecosystem. • Major changes in the Arctic regions with a substantial loss of Arctic sea-ice. • Major increase in extreme weather events and the spread of infectious disease. • Major increase in heat related morbidity and mortality, especially in the low latitudes. • Significant impacts on vulnerable ecosystems (polar regions, wetlands, cloud forests, and mangroves). • Significant increase in coastal and river flooding around the world. • Significant impacts on low latitude fisheries. • Decrease in crop yields and productivity especially in the tropics and sub-tropical regions.
3°-4°C	• Major impacts on all ecosystems including significant increase in species extinctions. • Loss of all warm water and many coldwater coral reef ecosystems. • Arctic completely free of sea ice in summer, Arctic temperature increase by 8°C. • Majority of mountain glaciers have disappeared, including all ice on Kilimanjaro (Tanzania). • Major increase in extreme weather events and spread of infectious disease. • Major decreases in agricultural and fishery production and available water resources. • Food and water security become major political and humanitarian issues. • Environmental forced mass migration increases. • Ocean and terrestrial carbon sinks reduce accelerating climate change.
4°- 5°C	• Catastrophic loss of ecosystems and species all around the world. • Melting of Western Antarctic/Greenland ice sheets accelerate, causes significant global sea level rises. • Fifth of world population affected by flooding and major coastal cities are abandoned. • Environmental forced mass migration accelerates and there is an increase in conflicts over resources. • In many countries summer temperatures persistently stay above 40°C. • Heat waves with temperatures as high as 50°C have become common. • Over 3.5 billion people are now water stressed. • Wildfires have created major air pollution events and human health crises. • Global food production plummets, leading to widespread malnutrition and starvation.
5°- >6°C	• Do not go there.

severity of heat waves, droughts, wild fires, storms and floods. Coastal cities and towns will be especially vulnerable as sea-level rises, increasing the impacts of floods and storm surges. Water and food security as well as public health will become the most critical problems facing all countries. Climate change threatens global biodiversity and the wellbeing of billions of people. In Table 4, potential impacts of climate change are summarised. Though many colleagues are planning how to deal with a increased 4°C world, my simple advice is, let us not go there.

Chapter 8

Climate Tipping Points

The composition of the atmosphere is changing beyond anything that has been experienced over the past three million years. Earth's climate is heading for unknown territory and therefore scientific uncertainty is enormous. The study of past climates reveals that the climate system can switch into a new state very quickly once a threshold has been passed. For example, ice-core records suggest that half the warming in Greenland at the end of the last ice age was achieved in only a few decades. This chapter examines the possibility that there are thresholds or tipping points in the climate system that may occur as we warm the planet.

Thresholds and Tipping Points

The relationship between a climate forcing factor such as GHGs and the climate response is complicated. In an ideal world, it would be a simple relationship with little or no delay, but we already know that there is inertia in the climate system, so that it responds to GHG forcing with a 10- to 20-year delay, depending on how much is being emitted. We can, therefore, examine the way different parts of the climate system respond to climate change with four scenarios (see Figure 46):

(a) *Linear but delayed response* (*Figure 46a*). In this case, the increase in GHGs produces a delayed but direct response in the climate system whose magnitude is in proportion to the additional forcing. This can be equated to pushing a car along a flat road. At first, nothing happens

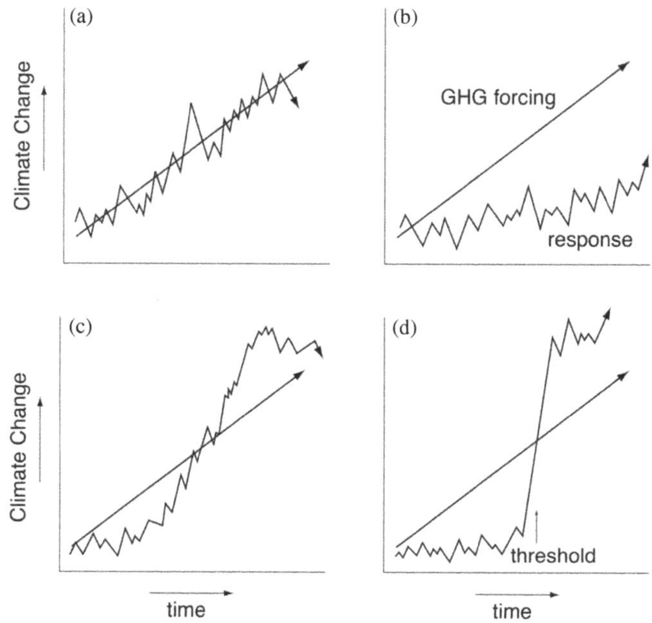

Figure 46. Response of the Climate System

as friction must be overcome before the car will start moving. Once the car is moving, most of the energy put into pushing is then used to move the car forward. An example of this is the heating up of the oceans, which has a delay effect due to the inertia of heating such a large body of water.

(b) *Muted or limited response* (*Figure 46b*). In this case, the GHG forcing may be strong, but the relevant part of the climate system is in some way buffered, and therefore gives very little response. This is analogous to pushing the car up a hill: you can spend as much energy as you like trying to push the car upwards, but it will not move very far. An example of this is the East Antarctic ice sheet, which has been stable at much warmer temperatures than today.

(c) *Delayed and non-linear response* (*Figure 46c*). In this case, the climate system may have an initial slow response to the GHG forcing but then responds in a non-linear way. This is a real possibility when it comes to climate change if we have underestimated the positive feedback in the system. This scenario can be equated to the car near the top of a hill: it takes some effort and thus time to push the car to the very top of the hill; this is the buffering effect. Once the car has reached the peak, it takes very little effort to push the car over it, and then it accelerates down the hill with or without help. Once it reaches the bottom, the car then continues for some time — the overshoot — and then it slows down of its own accord and settles into a new state.

(d) *Threshold response* (*Figure 46d*). In this case, initially, there is very little response in that part of the climate system to the GHG forcing; however, on reaching a threshold, all the response takes place within a very short period of time, in one large step. In many cases, the response may be much larger than one would expect from the size of the forcing and this can be referred to as a 'response overshoot'. This scenario equates to the bus hanging off the cliff at the end of the original film *The Italian Job*: as long as there are only very small changes, nothing happens at all. However, a critical point (in this case weight) is reached and the bus (and the gold) plunges off the cliff into the ravine below. An example of this could be the Greenland ice sheet that has started to melt; the melting could suddenly accelerate causing a catastrophic collapse.

There are also situations in which a threshold becomes a tipping point. You can think of a threshold as a point at which some changes in the system can still be reversed. However, a tipping point is a threshold that, when crossed, means the climate system moves into a new state and this transition is irreversible. An added complication when assessing whether climate change will create a simple threshold or a tipping point are bifurcations within the climate system. This means the forcing required to push the climate system one way across the threshold is different from the reverse. This implies that once a climate threshold has been passed, it is a lot more difficult to reverse it, and in some cases, it may in effect be irreversible.

The term 'tipping points' is used a lot in climate change research and discussions[1]. However, care must be taken as there are two usages of this word. First, there are references to climate tipping points, which are the large-scale, irreversible shifts in the climate system, such as irreversible melting of ice sheets or the release of huge stores of methane from below the oceans. The other usage concerns societal tipping points, which occur when climate change has a major effect on a region or a particular country. For example, a 200-mile (~322-km) shift northward of the South-East Asian monsoonal rainfall belt is, in climatological terms, a small shift, not a fundamental climate tipping point; but the same shift is a major 'social' climate tipping point for the countries where the rains no longer fall or those where it does for the first time, as their weather may have been permanently altered.

Climate Tipping Points

Scientist have identified nine global core tipping points and seven regional impact climate tipping points[2], see Figure 47. The nine global tipping points include: Greenland ice sheet collapse, Western Antarctic ice sheet collapse, Labrador Seas deep water circulation collapse, East Antarctic subglacial basins collapse, Arctic Winter Sea Ice collapse, Eastern Antarctic ice sheet collapse, Amazon rainforest dieback, high latitude permafrost collapse, and Atlantic deep water circulation collapse. Table 5 summarises the timeframe and the possible impact of these tipping points.

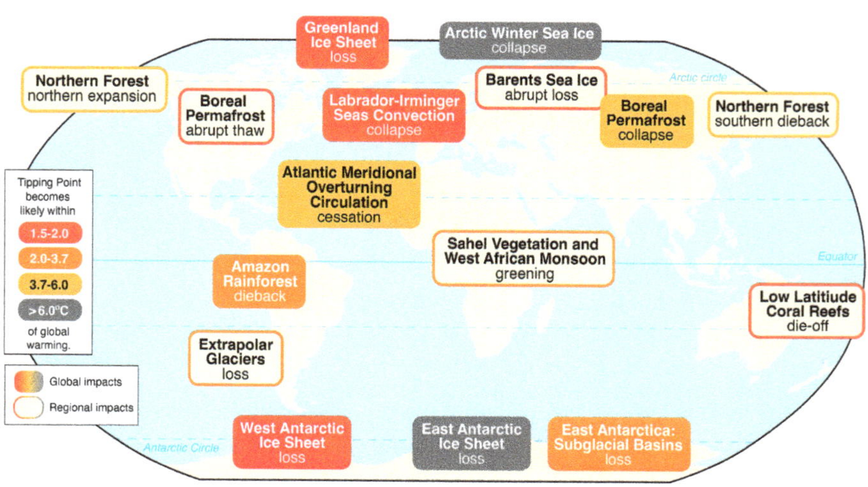

Figure 47. Tipping Points in the Climate System

Table 5. Global Tipping Points

Global core tipping elements								
Proposed climate tipping element (and tipping point)	**Threshold (°C)**			**Timescale (years)**			**Maximum Impact (°C)**	
	Estimated	**Minimum**	**Maximum**	**Estimated**	**Minimum**	**Maximum**	**Global**	**Regional**
Greenland Ice Sheet (collapse)	1.5	0.8	3.0	10,000	1,000	15,000	0.13	0.5 to 3.0
West Antarctic Ice Sheet (collapse)	1.5	1.0	3.0	2,000	500	13,000	0.05	1.0
Labrador-Irminger Seas/SPG Convection (collapse)	1.8	1.1	3.8	10	5	50	–0.5	–3.0
East Antarctic Subglacial Basins (collapse)	3.0	2.0	6.0	2,000	500	10,000	0.05	?
Arctic Winter Sea Ice (collapse)	6.3	4.5	8.7	20	10	100	0.6	0.6 to 1.2
East Antarctic Ice Sheet (collapse)	7.5	5.0	10.0	?	10,000	?	0.6	2.0
Amazon Rainforest (dieback)	3.5	2.0	6.0	100	50	200	0.1 (partial) 0.2 (total)	0.4 to 2.0
Boreal Permafrost (collapse)	4.0	3.0	6.0	50	10	300	0.2–0.4	~
Atlantic Meridional Overturning Circulation (collapse)	4.0	1.4	8.0	50	15	300	–0.5	–4 to –10

The regional tipping points include: low-latitude coral reefs die-off, abrupt thaw of the Boreal Permafrost, loss of Barents Sea ice, loss of mountain glaciers, greening of the Sahel and West Africa, Boreal forest northern expansion and Boreal forest southern die-off. Table 6 summarises the timeframe and the possible impact of these tipping points.

Out of these tipping points, current evidence suggests that one regional and three global climate tipping points could pass the critical threshold if global warming stays consistently above 1.5 °C. These are the Greenland ice sheet and West Antarctic ice sheet collapse (which combined, could contribute more than 12 m of global sea level rise), tropical coral reef die-off, and the abrupt thawing of the Boreal Permafrost. The other tipping points could occur beyond 2°C of global warming. In this chapter, we will focus on the irreversible melting of the Greenland and/or Western Antarctic ice sheet, slowing down of the North Atlantic deep ocean circulation, potential release of methane from melting gas hydrates, the Amazon rainforest dieback and the tropical coral reef die-off.

Melting Ice Sheets

The IPCC projections for sea-level rise by 2100, if there are no significant curbs to carbon emissions, are between 0.50 m and 1.3 m. The largest uncertainty within these estimates is the contribution that the melting of Greenland and Antarctica will make by the end of the century[3]. At the moment, it is estimated that Greenland is losing over 230 gigatonnes of ice per year, a seven-fold increase since the early 1990s. Antarctica is losing about 150 gigatonnes of ice per year, a five-fold increase since the early 1990s. Most of this loss is from the northern Antarctic Peninsula and the Amundsen Sea sector of West Antarctica. Greenland and Antarctica together constitute one of the most worrying potential climate surprises. If the large ice sheets there melted completely, their contribution to global sea-level rise would be as follows: Greenland, about 7 m; West Antarctic ice sheet, about 8.5 m; and East Antarctic ice sheet, about 65 m. These numbers are enormous when compare with just 0.3 m of sea-level rise if all the mountain glaciers melted. Paleoclimate data shows that the huge East Antarctic ice sheet developed 35 million years ago due to the

Table 6. Regional Tipping Points

Regional impact tipping elements								
Proposed climate tipping element (and tipping point)	**Threshold (°C)**			**Timescale (years)**			**Maximum Impact (°C)**	
	Estimated	**Minimum**	**Maximum**	**Estimated**	**Minimum**	**Maximum**	**Global**	**Regional**
Low-latitude Coral Reefs (dieoff)	1.5	1.0	2.0	10	~	~	~	~
Boreal Permafrost (abrupt thaw)	1.5	1.0	2.3	200	100	300	0.04 per °C by 2100; 0.11 per °C by 2300	~
Barents Sea Ice (abrupt loss)	1.6	1.5	1.7	25	?	?	~	+
Mountain Glaciers (loss)	2.0	1.5	3.0	200	50	1,000	0.08	+
Sahel and W.African Monsoon (greening)	2.8	2.0	3.5	50	10	500	~	+
Boreal Forest (southern dieoff)	4.0	1.4	5.0	100	50	?	net –0.18	–0.5 to –2
Boreal Forest (northern expansion)	4.0	1.5	7.2	100	40	?	net +0.14	0.5–1.0

progressive tectonic isolation of Antarctica, and that it has in fact remained stable in much warmer climates. Climate scientists, therefore, have a very high degree of confidence that the East Antarctic ice sheet will remain stable in this century as well.

Scientists are concerned that the melting of Greenland or the West Antarctic could significantly accelerate in the next 100 years. Even if we have already started the processes of melting the whole of these ice sheets, there is a physical constraint to the speed at which the ice can melt. This is due to the time it takes for heat to penetrate the ice sheets. Imagine dropping an ice cube into a hot cup of coffee. You know it will melt entirely, but it takes time for the heat to penetrate to the middle of the ice cube. Most of the ice from these ice sheets flows through ice streams to get to the sea, and there is a limit on how much ice these ice streams can transport. The worst-case scenario, according to leading glaciologists, is that these ice sheets could potentially add between 1 m to 1.5 m to the sea level by the end of the century, which would threaten many coastal populations around the world. There is also scientific debate about what happens to both the Greenland and Antarctic ice sheets beyond the next 100 years; even if significant melting does not occur this century, we may have started a process that causes irreversible melting during the next one. Our carbon emission over the next few decades could determine the long-term future of the ice sheets and the livelihoods of billions of people who live near the coast.

Changing Ocean Circulation

The circulation of the ocean is one of the major controls on our global climate. As discussed in Chapter 4, the deep ocean is the only candidate for driving and sustaining internal long-term climate change (of hundreds to thousands of years) because of its volume, heat capacity, and inertia. In the North Atlantic, the north-east trending Gulf Stream carries warm and salty surface water from the Gulf of Mexico up to the Nordic seas. The increased saltiness, or salinity, in the Gulf Stream is due to the huge amount of evaporation that occurs in the Caribbean, which removes moisture from the surface waters and concentrates the salt in the sea water. As the Gulf Stream flows northward, it cools down. The combination of a high salt content and low temperature increases the density of the surface water. Hence, when it reaches the relatively low salinity oceans north of Iceland and in the Labrador Sea, the surface water has cooled sufficiently to become dense enough to sink into the deep ocean. The 'pull' exerted by the sinking of this dense water mass helps maintain the strength of the warm Gulf Stream, ensuring that a current of warm tropical water continuously flow into the north-east Atlantic, sending mild air masses across to the European continent. This impact is apparent when comparing the climate on either side of the Atlantic Ocean. Think of the climate of these areas which are at the same latitude, London with Labrador, or Lisbon with New York. Or a better comparison is between Western Europe and the west coast of North America, which have a similar geographical relationship between the ocean and continent — so think of Alaska and Scotland, which are at about the same latitude.

The newly formed deep water sinks to a depth of between 2,000 m and 3,500 m in the ocean and flows southward down the Atlantic Ocean, as the North Atlantic Deep Water (NADW). In the South Atlantic Ocean, it meets a second type of deep water, which is formed in the Southern Ocean and is called the Antarctic Bottom Water (AABW). This is formed in a different way compared to NADW. Antarctica is surrounded by sea ice and its deep water is formed in coast polynyas, or large holes in the sea ice. Out-blowing Antarctic winds push sea ice away from the continental edge to produce these holes. The winds are so cold that they super-cool the exposed surface waters. This leads to more sea-ice formation and salt rejection, producing the coldest and saltiest water in the world. AABW flows around the Antarctic and penetrates the North Atlantic, flowing under the warmer, and thus somewhat lighter, NADW (see Figure 48). The AABW also flows into both the Indian and Pacific Oceans.

This balance between the NADW and AABW is extremely important in maintaining our present climate, as not only does it keep the Gulf Stream flowing past Europe, but it also maintains the right amount of heat exchange between the Northern and Southern Hemispheres. Scientists have shown that the circulation of deep water can be weakened if there is sufficient input of fresh water to make the surface water too light to sink. This evidence has come from both computer models and the study of past climates. Scientists have coined the phrase 'dedensification' to mean the removal of density by the addition of fresh water and/or warming of the water, both of which prevent sea water from being dense enough to sink. As we have discovered, there is already concern that global warming will cause significant melting of the polar ice caps. This will lead to more fresh water being added to the polar oceans. Climate change could, therefore, cause the weakening of the NADW, and a weakening of the warm Gulf Stream (Figure 48b). This would cause much colder European winters and

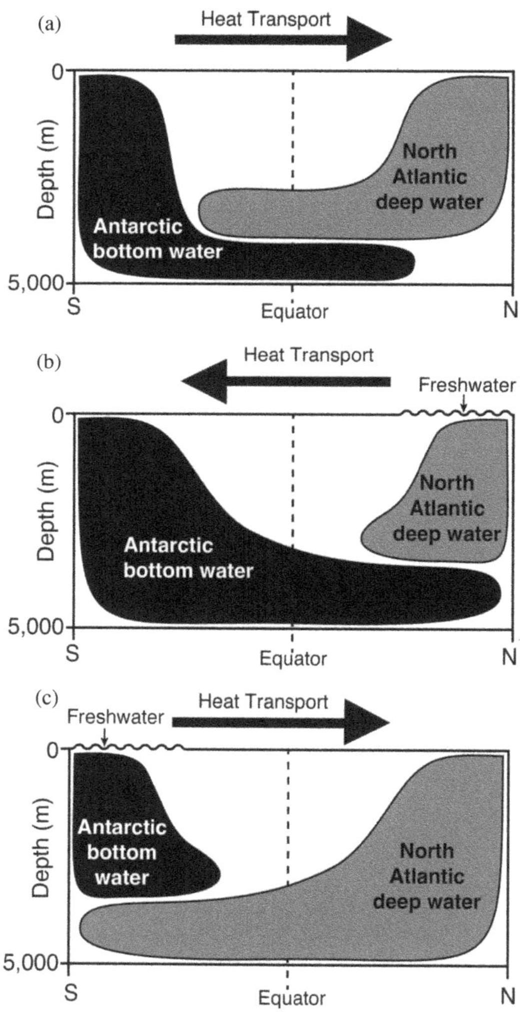

Figure 48. Deep Ocean Circulation Changes Depending on Freshwater Inputs

more severe weather. However, the influence of the warm Gulf Stream is mainly seen in the winter and has only a small effect on summer temperatures. So, if the Gulf Stream weakens, global warming would still cause European summers to heat up. Europe would end up with extreme seasonal weather, very similar to that of Alaska[3].

It has been over 40 years since the possibility of a catastrophic shut down of the deep ocean circulation was suggested, and there has been a huge amount of work on it. Monitoring has shown that the Gulf Stream has weakened by 15% since the middle of the last century. Evidence collated from this ocean monitoring and climate model predictions of the future in the very latest IPCC report suggest that a significant weakening of the Gulf Stream is unlikely in the 21st century. Some models do, however, show a significant weakening in the overturning of the North Atlantic in this century, especially in the high emission scenarios, and the problem is that we do not know where a potential tipping point leading to changes in the deep ocean circulation might be. If Greenland or Western Antarctica melting accelerates, then increased amounts of freshwater could enter the oceans, significantly disrupting the deep ocean system.

Gas Hydrates

A large amount of carbon in the form of methane is stored beneath the world's oceans. The methane gas is trapped in a solid cage of water molecules at low temperatures and/or high pressures. The methane gas comes from decaying organic matter found deep in ocean sediments and in soils beneath permafrost. These gas hydrate reservoirs could be unstable, as an increase in temperature or decrease in pressure would cause them to destabilise and to release the trapped methane. Climate change is warming up both the oceans and the permafrost, threatening the stability of gas hydrates[4]. Methane is a strong GHG and is up to 30 times more powerful than carbon dioxide (see Table 1). If enough were released, it would raise global temperatures, which could lead to the release of even more gas hydrates — producing a runaway effect. Scientists really have no idea how much methane is stored in the gas hydrates beneath our feet: estimates are between 1,000 and 10,000 gigatonnes (compared with ~800 gigatonnes of carbon currently in the atmosphere), which is a huge range.

Without a more precise estimate, it is very difficult to assess the risk posed by gas hydrates.

The reason why scientists are so worried about this issue is because there is evidence that a super greenhouse effect occurred 55 million years ago, during what is called the Palaeocene–Eocene Thermal Maximum (PETM)[5]. During this hot-house event, scientists estimated that up to 1,500 gigatonnes of gas hydrates might have been released. This huge injection of methane into the atmosphere accelerated the natural greenhouse effect, producing an extra 5°C of warming. There is still, however, considerable debate over the PETM. For example, some colleagues have suggested that the warming was caused by carbon dioxide released from a phase of massive volcanism instead of a massive gas hydrate methane release.

The current consensus is that the ocean reserves of gas hydrate are likely to remain stable this century. Gas hydrates form a solid layer at the bottom of the ocean. The depth of this layer is controlled by the geothermal heat gradient — as you go deeper in the sediment, the temperature rises at about 30°C/km. At a certain depth, it is too warm for gas hydrates to exist and methane collects there as free gas in the sediment. As ocean temperatures change, the temperature change has to be transmitted through the solid gas hydrate layer to the lower boundary for some of it to melt. If this process is slow enough, the gas released migrates upwards in the ocean sediment column and re-freezes at a higher level. However, if carbon emissions are not curbed, we could see this process speed up by the next century, leading to the release of some of the methane stored in the deep ocean.

It is clear that the gas hydrate below what was once permafrost is already melting, with bubbles observed in many Canadian and Siberian lakes. With the Arctic amplification, temperature rises will be nearly twice the global average in the northern polar regions which will accelerate the gas hydrate melting. But we still do not have an indication of how much methane is stored beneath the world's permafrost regions. So, at the moment, our best estimate suggests a global warming of 3°C could release between 35 and 940 gigatonnes of carbon, which could add between 0.02°C to 0.5°C to global temperatures.

Amazon Dieback

The Amazon River discharges approximately 20% of all fresh water carried to the oceans. The Amazon drainage basin is the world's largest, covering an area of 7,050,000 km^2, about the size of Europe. The river is a product of the Amazon monsoon, which every summer brings huge rains. This also produces the spectacular expanse of rainforest, which supports the greatest diversity and the largest number of species of any area in the world.

The Amazon rainforest is important when it comes to climate change as it is a huge natural store of carbon. Originally, it was thought that established rainforests such as the Amazon had reached maturity. Detail surveys of all the rainforests of the world over the past four decades show evidence that this is incorrect. In the 1990s, intact tropical forests — those unaffected by logging or fires — removed roughly 46 billion tonnes of carbon dioxide from the atmosphere. The sting in the tail is that this removal had diminished to an estimated 25 billion tonnes in the 2010s. The lost sink capacity is 21 billion tonnes of carbon dioxide, equivalent to a decade of fossil fuel emissions from the UK, Germany, France, and Canada combined. All this data is compiled by the African Tropical Rainforest Observatory Network and the Amazon Rainforest Inventory Network. Over 300,000 trees are being tracked, with more than 1 million diameter measurements in 17 countries, and the data is standardised and managed by the University of Leeds[6].

The concern about a possible Amazon rainforest dieback came from two seminal papers published in 2004 by colleagues at the UK Meteorological Office's Hadley Centre[7]. Their climate model was the first to include vegetation–climate feedback and suggests that global warming by 2050 could increase the winter dry season in Amazonia. For the Amazon rainforest to survive, it requires not only a large amount of rain during the wet season but a relatively short dry season so as not to dry out. Climate models suggest climate change could cause the global climate to shift towards a more El Niño-like state with a much longer South American dry season or more frequent extreme long droughts[8]. The Amazon rainforest cannot survive these drier conditions and would be

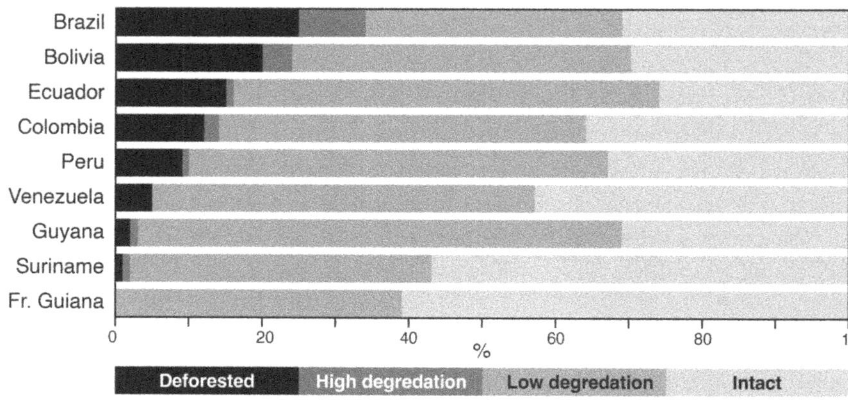

Figure 49. Amazon Forest Deforestation and Degradation

replaced by savannah (dry grassland), which is found both to the east and south of the Amazon basin today (Figure 49). This replacement would occur because the extended dry periods would lead to forest fires destroying large parts of the rainforest. This is exactly what was seen during the extreme droughts of 2005, 2010, 2015/16, and 2023. The Amazon 2023 drought was the most severe in at least 45 years. The wildfires also return the carbon stored in the rainforest back into the atmosphere, accelerating climate change. The savannah would then take over those burnt areas, as it is adapted to coping with the long dry season. Unfortunately, savannah has a much lower carbon storage potential per square kilometre compared to the rainforest.

Modeling the Amazon forest response to climate change is complicated because there are positive and negative feedbacks. For example, higher levels of atmospheric carbon dioxide have a "fertilisation" effect on plants and trees, boosting photosynthesis and promoting growth. It also makes plants more water efficient and hence more drought tolerant, offsetting some of the effects of the longer predicted dry season. Other climate models have not found such a profound dieback and the current IPCC review suggests a sustained dieback of the Amazon rainforest is unlikely this century — so long as the Amazon rainforest stays intact. It is that last part which is the biggest issue, because under the leadership of Brazilian President Jair Bolsonaro, deforestation rates did rise significantly, accompanied by a significant increase in forest fires, many occurring in areas which do not usually suffer from them, indicating that many are being started deliberately. Deforestation rates in the Amazon have now decreased but it is still happening, just at a slower rate. The deforestation and fragmentation of the Amazon and other rainforests around the world make them more vulnerable to climate change, and hence make the likelihood of a catastrophic dieback more likely.

Coral Reef Die-Off

Over 500 million people around the world depend on coral reefs for food, income, tourism and coastal protection. Increased sea surface temperature over the last 50 years, due to climate change, has caused mass bleaching of coral, especially in sub-tropical regions[9]. Evidence shows that a sustained ocean temperature spike of just 1°C above average is enough to cause bleaching. Bleaching occurs when the coral gets heat stress and expels the algae which create the bright colours of the corals, turning them white. The photosynthetic algae, called zooxanthellae, have a symbiotic relationship with coral providing food while the corals provide them with shelter. Without the algae, the corals slowly die. Once the coral reefs die, many areas can become dominated by seaweed, making it very difficult for the corals to regrow without human intervention. Sea surface temperatures have already risen to 1.5 °C above pre-industrial times and the IPCC projected this will cause a further 70–90% coral reef damage. If the world warms by 2 °C, coral reefs would become extremely rare (see Table 4).

Summary

Human-induced climate change has already affected our planet and could have even more radical impact over the next 75 years. Scientists worry about potential surprises in the global climate system that could exacerbate future climate change[10]. As discussed above, these include the possibility that Greenland and/or the Antarctic could start to melt irreversibly, raising sea level by many metres in the next century. The North-Atlantic-driven deep ocean circulation could change, producing extreme seasonal weather in Europe. The Amazon rainforest could start to dieback due to the combined effects of deforestation and climate change, causing the loss of huge amounts of biodiversity and increasing carbon emissions to the atmosphere, driving further global warming. Finally, there is the threat of additional methane being released from gas hydrates beneath the oceans and permafrost, which could accelerate climate change. One way to ensure we avoid the worst effects of climate change and greatly reduce the likelihood of climate surprise is to keep climate change as small as possible. The aspiration of our global leaders is to try to keep climate change to just 1.5°C above pre-industrial levels. In the next chapter we explore how they came to this decision and how they hope to achieve it.

Chapter 9

The Politics of Climate Change

Introduction

The global politics of climate change has been driven by three factors: first is the science, second has been public pressure, and the third is the political realisation of the economic threat of climate change. As the science of climate change has been discussed earlier in this book, this chapter analyses how public pressure came about and what the economic implications are of not dealing with climate change. The chapter will also examine the role of the media and the fossil fuel companies in delaying meaningful action on climate change. First though, we need to understand which countries and regions have produced GHG emissions in the past and now.

Who Produces the Pollution?

The United Nations Framework Convention on Climate Change (UNFCCC) was created to produce international agreements on reducing global GHG emissions. This is not a simple task as CO_2 emissions are not evenly produced by countries. According to the Intergovernmental Panel on Climate Change, the primary source of CO_2 is the burning of fossil fuels; over 85% of the global CO_2 emissions comes from energy production, industrial processes, and transport[1]. These are not evenly distributed around the world because of the unequal distribution of industry and wealth; North America, Europe, and Asia emit over 90% of the global, industrially produced CO_2 (see Figure 50). Moreover, the developed nations have historically emitted much more than less developed countries[1].

The second major source, accounting for 10–15% of global CO_2 emissions, is land-use changes[1]. These emissions come primarily from the cutting down of forests for the purposes of agriculture, urbanisation, or roads. When rainforest is cut down, the land often turns into less productive grassland with considerably reduced capacity for storing CO_2. In land-use changes, the pattern of CO_2 emissions is different, with South America, Asia, and Africa being responsible for over 90% of present-day land-use change emissions. This raises important ethical questions because it is difficult to tell these countries to stop deforestation when this has already occurred in much of North America and Europe before the beginning of the 20th century. Besides, in terms of the amount of CO_2 released, industrial processes still significantly outweigh land-use changes.

We have put nearly half a trillion tonnes of carbon into the atmosphere since the Industrial Revolution, but this amounts to only half of our total emissions. The other half has been absorbed by the Earth; with 25% going

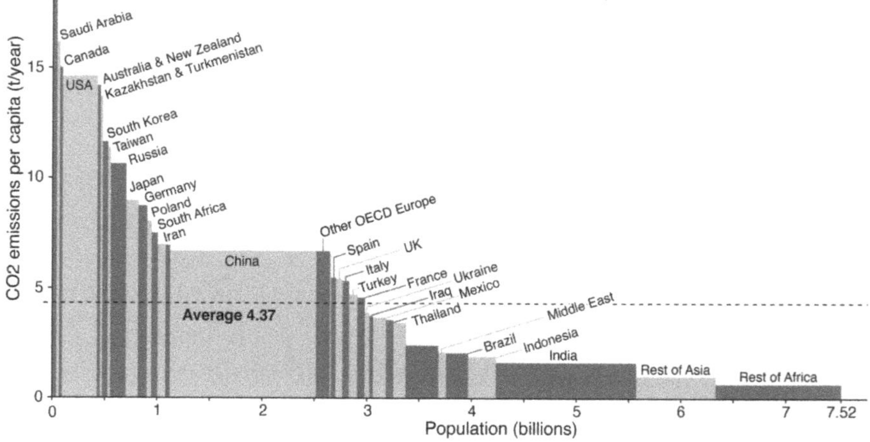

Figure 50. Historic Carbon Dioxide Emissions by Region

into the oceans and 25% going into the land biosphere. Scientists are concerned that this uptake of our pollution is unlikely to continue at the same level in the future. This is because as global temperatures rise, the warmer oceans will hold less dissolved CO_2. As we continue to deforest and convert land for farming and urbanisation, there will be less vegetation to absorb CO_2, again reducing the uptake of our carbon pollution.

The Rise of the Global Environmental Social Movement

There have been four main waves of environmental social movements. The first was in 1970s represented by the response to Rachel Carson seminal book 'Silent Spring' (1962). Focused on pollution, pesticides and destruction of local environments, many of the key environmental NGOs were formed around this movement, for example, Friends of the Earth (1969), Greenpeace (1971) and WWF (1961). The second wave sprung up in the late 1980s and early 1990s when a growing realisation emerged that humanity's environmental impact was global.

The 1980s saw a massive grass-roots expansion in the environmental movement, particularly in the USA, Canada, and the UK, partly as a backlash against the right-wing governments of the 1980s and the expansion of the consumer economy, and partly because of the increasing number of environment-related stories in the media. This heralded a new era of global environmental awareness and the emergence of transnational NGOs. The roots of this growing environmental awareness can be traced back to a number of key markers: these include the publication of Rachel Carson's Silent Spring in 1962; the image of Earth seen from the Moon in 1969; the Club of Rome's 1972 report on Limits to Growth; the Three Mile Island nuclear reactor accident in 1979; the nuclear accident at Chernobyl in 1986; and the Exxon Valdez oil spillage in 1989. However, these environmental problems were all regional in effect, limited geographically to the specific areas in which they occurred[2].

It was the discovery in 1985 by the British Antarctic Survey of the depletion of ozone over Antarctica that demonstrated the global connectivity of our environment[2]. The ozone 'hole' also had a tangible international cause, the use of chlorofluorocarbons (CFCs), which led to a whole new area of politics: the international management of the environment. There followed a set of key agreements: the 1985 Vienna Convention for the Protection of the Ozone Layer; the 1987 Montreal Protocol on Substances that Deplete the Ozone Layer; and the 1990 London and 1992 Copenhagen Adjustments and Amendments to the Protocol. These have been held up as examples of successful environmental diplomacy.

These new global environmental concerns and the ability to deal with them internationally was encouraged and articulated by the leading politicians at the time. For example, Margaret Thatcher, the Prime Minster of the UK in 1989, gave an address to the UN outlining the science of climate change, the threat it posed to all nations, and the action needed to advert the crisis[3]. World leaders agreed to set up the Intergovernmental Panel on Climate Change (IPCC) in 1988 and produced its very first science report in 1990[4]. Two years later, with support from leaders from all around the world, the UN held the Rio Earth Summit, officially called the United Nations Conference on Environment and Development (UNCED) to help member states cooperate on sustainability and protecting the world's environment[4].

The Economists Wade In

Economists have been involved with studying climate change from the very beginning of the IPCC process. The most important study was UK government-commissioned Stern Report on The Economics of Climate Change in 2006 (published in 2007)[5]. The report was led by Sir Nicholas Stern, then the adviser to the UK government on the economics of climate change and development reporting to Prime Minister Tony Blair. The report states that if we do nothing, the impacts of climate change could cost between 5% and 20% of world GDP every year by 2050. Given that the World economy could be twice as big by 2050, this would represent between $10 trillion and $40 trillion — for context, current GDP of China is just under $20 trillion. That means the whole world loses one-fifth of what it earns to address the impacts of climate change (discussed in Chapter 7). This of course puts climate change impacts on a completely different economic scale than was envisaged by previous studies.

Nonetheless, the Stern Report also present some good news: if we do everything that we can to reduce global GHG emissions and ensure we adapt to the coming effects of climate change, it will cost us only 1% of world GDP every year. The Stern Report was criticised by other economists — for example, does it use the right inherent discount rate? This is the rate economists use to take into account that consumption inherently has a lower value in the future than in the present. In other words, future consumption should be discounted simply because it takes place in the future and people generally prefer the present to the future. For example, the Nobel laureate William Nordhaus used inherent discount rates of up to 3%, which means that people today value environmental benefit occurring 25 years in the future half as much as the same benefit today. Nordhaus, however, came under intense criticism as he claimed a 4°C increase in global temperature over pre-industrial levels would only reduce GDP per capita by between 2% and 4%. But the fundamental flaw in this model is that he uses a linear, and not a quadratic damage function — so even catastrophic levels of climate change do not do much economic damage in his economic model. The Stern Report has also been criticised for being overly optimistic about the costs of adapting to a low-carbon world. In June 2008, Sir Nicholas Stern did revise his estimated costs up to 2% of world GDP. But counter to this, the detailed 'Project Downdraw' estimated that transiting to zero overall GHG emissions could save the World over $46 trillions[6]. The Stern Report sent seismic waves around the world. It was as if people said to themselves, 'if the economists are worried about the cost of climate change, it must be real'.

This was not the end of the involvement of economists in climate change and there have been a number of highly influential books and papers which question our whole understanding of economics and its relationship with the environment. These include the economist Tim Jackson's book 'Prosperity Without Growth', first published in 2009, that challenges the orthodox view that economic growth is required or even desirable[7]. In 2017, economist Kate Raworth published 'Doughnut Economics'[8], in which she shows seven ways classical economics has got it wrong, arguing that environmental limits and basic human rights must be at the centre of economics. For the first time for two generations, 'classical economics' is under sustained attack from a new generation of dynamic innovative 21st century economists who see environmental and human wellbeing as integral parts of the world economy. At the centre of this is how we deal with climate change while improving peoples' lives.

Climate Change and the Media

The other reason for the emergence of climate change as a major global issue was the intense media interest. This is because climate change is perfect for the media: a dramatic story about the end of the world as we know it with key protagonists arguing that it is not even real. The majority of newspaper articles the UK, USA and Australia in the 1990s cast doubt on the claims of climate change. There was a recurrent attempt to promote mistrust in science, through strategies of generalisation, of exaggerating disagreement within the scientific community, and, most importantly, discrediting scientists and scientific institutions[9].

There are two reasons for this extraordinarily media-facilitated public scientific debate. First, climate change deniers and industrial lobby groups who do not want to see political action to address climate change are using this debate about methods and scientific uncertainty as a convenient hook on which to hang their case for delay. In fact, it has been found that in 2019, five of the largest public listed oil companies spent over $200 million lobbying to control, delay, or block binding climate policy.

Secondly, the media's ethical commitment to balanced reporting, inappropriately applied, draws unwarranted attention to critical views when they are marginal and outside the realm of what is normally considered 'good' science. In the UK, the BBC came under increasing criticism for continually presenting this false balance, usually pitching a climate scientist against a seasoned politician or a paid lobbyist[10].

Beyond conventional media, the so-called debate about climate change has moved on to social media, with climate change deniers attacking the evidence and views of scientists whenever they can. This rise of fake news has impacted many areas of science including vaccinations and efforts to tackle Covid-19, as well as climate change. Together, false balance in media debate, fake news, and social media campaigns contribute to a public impression that the science of climate change is 'contested', despite what many would argue is an overwhelming scientific case that climate change is occurring and that human activity is its main driver[11]. Climate change denial and the false narratives have surged again on social media and in mainstream politics since the election of President Trump in the USA. Fortunately, these seem to have had little effect on the public and there is significant proportion of people who have realised that climate change is real and is a major threat. This is mainly through people's own experience, listening to experts and influencers, seminal documentaries and/or watching the effects of extreme weather around the world. United Nations Development Programme (UNDP) and University of Oxford's 2024 People's Climate Vote found that 80% of people worldwide want stronger government climate action[12]. Now this huge survey statistically represents 87% of the world's population. There were other important findings, for example: 86% agreed that countries should set aside geopolitical differences, such as those regarding trade and security, and work together on climate change. Worldwide, 72% support a quick transition away from fossil fuels to renewable sources of energy. Even in the world's biggest producers of fossil fuels, the majority of people support a quick transition away from coal, oil and gas. Last but not least, four in five people (79%) expressed support for climate justice, agreeing that rich countries should increase the help given to poorer countries to address climate change and its impacts.

The New Global Environmental Social Movements

In 2008 and 2009, there was a third global rise in the social awareness of climate change. This time, it was focused on the hope of a major climate deal at the Copenhagen climate conference. Sadly, Copenhagen ended in abject failure due to the lack of international leadership, sabotage by the US and the global worries about dealing with the 2008 global financial crash (see Chapter 10 for more details). It took until the Paris 2015 climate conference to get the negotiation back on track. For almost 10 years, the environmental movement was held back due to the focus on the global economy. This all changed in 2018, when the fourth wave of the global environmental social movement began[13].

In May 2018, the protest group Extinction Rebellion was set up in the UK and launched in October 2018 with over 100 academics calling for action on climate change. The aim of Extinction Rebellion is to use non-violent civil disobedience to compel Governments around the world to avoid tipping points in the climate system and biodiversity loss to avoid both social and ecological collapse. In November 2018 and April 2019, they brought central London to a standstill, and Extinction Rebellion has now spread to at least 60 other cities around the world.

In August 2018, the 15-year-old Greta Thunberg started to spend her school days outside the Swedish parliament holding a sign saying Skolstrejk för klimatet (School strike for climate), calling for stronger action on climate change. The message spread. Soon, other students all around the world started similar school strikes once a month on a Friday, and they called the movement 'Fridays for Future'. It has been estimated that by the end of 2019, there had been over 4500 strikes across over 150 countries, involving 4 million school children.

In 2018 and 2019, three extremely influential IPCC reports were published. The first, in 2018, was the Special Report on Global Warming of 1.5°C, which documented what the world need to do if global temperature rise was to be kept at only 1.5°C[14]. It also showed the positive and negative interactions of climate change mitigation and the sustainability development goals. The second was the IPCC special report on the land, and how climate change would impact desertification, land management, food security, and the terrestrial ecosystems[15]. The third was the IPCC Special Report on the Ocean and Cryosphere[16], showing the impacts of climate change on the speed of melting of ice sheets, mountain glaciers and sea ice, and their implications of sea level rise and marine ecosystems.

There is a school of thought that we have now entered a fifth wave of social movement — radical direct action. Borne out of the frustration that many climate campaigners feel from the lack of action by Government and many corporations, direct actions have included: Insulate Britain protestors gluing themselves to motorways; Just Stop Oil protestors throwing paint over famous painting; protestors letting down the tyres of large SUVs in cities; and the sabotaging of oil pipelines and refineries. Many activists now see direct action and violence as the only way for government authorities to take note and act — in the US, the FBI has defined these actions as eco-terrorism.

This new social movement, and the very latest science inspired many corporations to take a leading role. Microsoft has set the agenda for the technology sector with the ambitious target to go carbon negative by 2030. By 2050, they want to remove all the carbon pollution from the atmosphere that they and their supply chain have emitted since the founding of the company in 1975. Microsoft is just one of many companies taking action. For example, over 7,000 businesses world-wide have pledged to adopt science-based targets, with 5,500 already having emission reduction targets validated by the initiative. Science based targets effectively mean they will all be net zero carbon emissions by 2050.

Given all this pressure, governments all around the world in 2019 started to declare that we are in fact, in a climate emergency and that action must be taken. At the time of the publication of this book, over 2,400 jurisdictions across 40 countries have declared a climate emergency, a declaration that covers over 1 billion people globally. This is despite the fact that in many countries, there is a backlash against climate action due to political expediency and fossil fuel industry lobbying.

Summary

The science of climate change has made us examine the whole basis of modern society. It has set politicians arguing, set nations against each other, queries individual choices of lifestyle, and ultimately asks questions about humanity's relationship with the rest of the planet. Over the last 30 years, there has been intense media and political discussion about climate change and how to deal with it. On one side are those wanting to maintain the status quo — as they are making billions of dollars out of the fossil fuel industry, while on the other side are those wanting a rapid transition away from fossil fuels. Climate change has become deeply political, and arguments are rarely about the science — in many cases, the science is simply denied and ignored. Climate change has become a cultural battle ground and allegiance to political dogma has become more important than evidence and facts.

Chapter 10

International Climate Negotiations

Introduction

The most logical approach to the climate change problem is to significantly cut GHG emissions. At the Paris climate meeting in 2015, world leaders agreed that global temperature increase should be kept below 2°C, with an aspirational target of 1.5°C. At subsequent climate meetings, world leaders have ratified 1.5°C as the international agreed target. Despite this agreement, global carbon emissions have continued to rise every year. The only exception was 2020 when the global lockdown prompted by the Covid-19 pandemic dropped emissions by about 5%. Ceasing almost all flights and car journeys around the world had a small impact on our total greenhouse gas pollution. In fact, global carbon emissions for 2020 with a global pandemic were the same as 2006. This was because there was very little change to energy production during the pandemic; and subsequently, GHG emission dip has gone away and the highest ever GHG emission occurred 2025. This chapter examines the successes and failures of the international climate negotiations — and why despite all these efforts, GHGs are still rising.

Climate Change Negotiations

The last 30 years have been a period of intense and continuous international negotiation to deal with climate change. During the same 30 years, humanity has doubled the amount of anthropogenic carbon dioxide in the

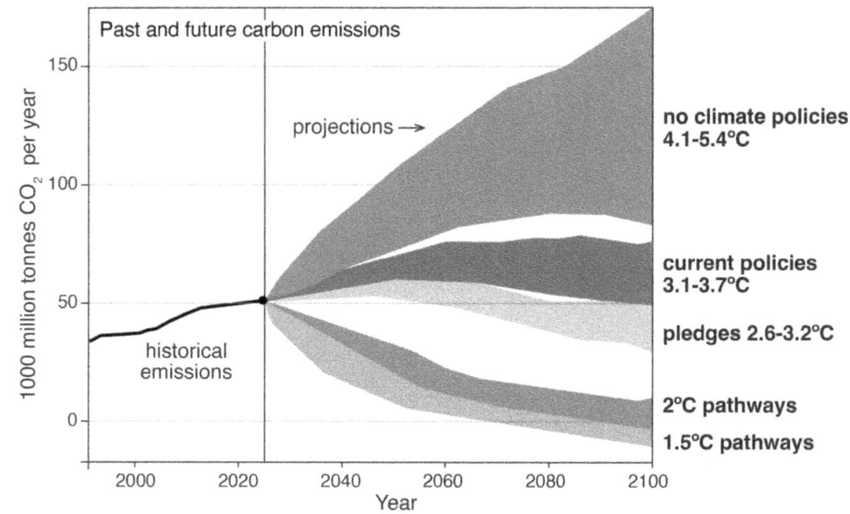

Figure 51. Potential Future Global Warming Based on Different Carbon Emissions

atmosphere. There has, however, been progress and some notable successes. In 2015 at COP21, 196 countries signed the Paris Agreement stating that they would limit global temperatures to well below 2°C above pre-industrial levels and would pursue efforts to limit the temperature increase to 1.5°C above pre-industrial levels[1]. The first review of the Paris agreement and a critical stock take was at COP26 in Glasgow with many countries pledging to go net zero emissions by the middle of the century. But currently, these pledges, if fulfilled, will only limit global average temperature to 2.4°C to 2.8°C (Figure 51). At COP27 in Egypt, the core agreements from the Glasgow Climate Pact were kept and countries finally agreed to set up a Loss and Damage facility[2] — though details of who finances and who can claim are still be to be worked out.

In the Beginning

In 1989, Margaret Thatcher, the Prime Minster of the UK, gave an address to the UN outlining the science of climate change, the threat it posed to all nations, and the actions needed to avert the crisis[3]. She summed up by saying: "We should work through this great organisation and its agencies to secure world-wide agreements on ways to cope with the effects of climate change, the thinning of the ozone layer, and the loss of precious species." This sentiment was echoed in similar speeches by George Bush Senior, President of the United States, including one in 1992 when he outlined his 'Clear Skies' and 'Global Climate Change' initiatives at the National Oceanic and Atmospheric Administration.

The Intergovernmental Panel on Climate Change (IPCC) was set up in 1988 and produced its very first science report in 1990. Two years later, with support from leaders from all around the world, the UN held the Rio Earth Summit — officially called the United Nations Conference on Environment and Development (UNCED) — to help member states cooperate on sustainability and protect the world's environment. The Summit was a huge success and led to the Rio Declaration on Environment and Development, the local sustainability initiative called Agenda 21 and Forest Principles. Following negotiations at UN General Assembly in December 1990, the Framework Convention on Climate Change (UNFCCC) that underlies the negotiations to limit global greenhouse gas emissions was open for parties to sign up to at UNCED. The conferences also allowed the parties to join the United Nations Convention to Combat Desertification and the Convention on Biological Diversity. The Rio Earth Summit also laid the foundations for the Millennium Development Goals and the subsequent Sustainable Development Goals[1].

The United Nations Framework Convention on Climate Change (UNFCCC) came into force on 21 March 1994. As of August 2022, the UNFCCC has 196 parties. The UNFCCC currently operates by "agreement by consensus". Decisions are taken by consensus only because there has been no agreement on the draft rules of procedure and the default is consensus — which of course allows any country to block any decision[4]. Due to the different levels of economic development between countries, the UNFCCC originally recognised two different groups: Annex I, which include all the developed countries, and non-Annex I, which include the less developed and rapidly developing countries[4]. Annex I was subsequently divided when some countries argued that their economies were in transition. As a result, the richest countries were placed in an additional category, Annex II. With the Paris Agreement, the UNFCCC has moved away from the strict bifurcation of Annex I/Non-Annex I and has a more fluid approach instead of emphasising "developed country Parties" and "developing country Parties". At the outset of the UNFCCC process was

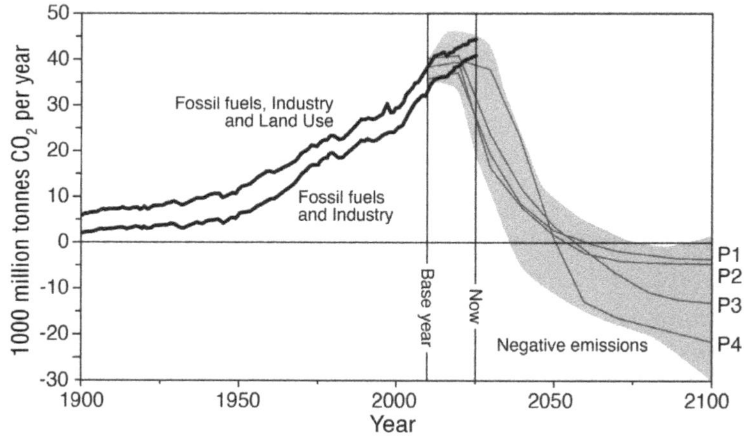

Figure 52. Pathways to Achieving a 1.5°C Stable World

the principle of common but differentiated responsibilities. This was because the UNFCCC acknowledged that different countries have historically emitted varying quantities of GHGs, and therefore need to make greater or lesser efforts to reduce their emissions. For example, if emissions from both fossil fuel use and land use changes are considered then from 1850 to 2026, the USA emitted over 20% of global cumulative CO_2 emissions compared to China at 12% and India at 4%. Today, annual per capita emissions of CO_2 in the USA are about ten times greater than in India. With the Paris Agreement, this approach shifted to the idea of 'contraction and convergence' — the idea that every country must reduce its emissions and that all countries must converge on net zero emissions[5]. The importance of a net zero emissions target was agreed in the 2015 Paris agreement and the seminal IPCC report on Global Warming of 1.5°C published in 2018 showed that to achieve the target of 1.5°C, there had to be zero CO2 emissions by about 2050, and then negative CO2 emissions for the rest of the century (Figure 52).

COP3 Kyoto 1997: The First International Treaty

Since the UNFCCC was set up, every nation, better known as 'the parties', have met annually at the 'Conference of the Parties' (COP) to move negotiations forward. Only five years after the UNFCCC was created, at COP3 in December 1997, the first international agreement to cut GHG emissions was drawn up, named the Kyoto Protocol. This restated the UNFCCC general principles for a worldwide treaty on cutting GHG emissions and, more specifically, that all developed nations would aim to cut their emissions by 5.2% relative to 1990 levels by 2008–12. The USA, under the leadership of Al Gore, signed the Kyoto Protocol in 1998, but when George Bush took over as President, the USA did not ratify the Kyoto Protocol at Bonn in 2001. With the USA producing about one-quarter of the world's CO2 pollution at the time, this was a big blow for the treaty. Moreover, the targets set by the Kyoto Protocol were made more flexible during the Bonn meeting to ensure that Japan, Canada, and Australia would join. Australia finally made the Kyoto Protocol legally binding in December 2007.

To balance out the historic legacy of emissions by developed countries, the treaty did not include developing countries, but it was assumed that developing countries would eventually join the post-2012 agreement. The Kyoto Protocol came into force in February 2005, after Russia ratified the treaty, thereby meeting the requirement that at least 55 countries representing more than 55% of the global emissions were participating.

COP8 to COP13: The Technical Years

COPs between 2002 and 2005 focused on some of the technical aspects of carbon accounting, carbon trading (the clean development mechanism) and the rules of the Kyoto protocol. The first COP at which the Kyoto protocol was legally in force was COP11 in Montreal, Canada, in 2005. After the protocol came into effect, COPs became formally COP/MOPs (meetings of the parties) so Montreal was also MOP1. Because the USA continued to stand outside the Kyoto protocol, a twin-track approach to the negotiations was adopted, whereby the COP countries that had ratified the protocol had a series of meetings separate from countries — chiefly the USA and Australia, among the major economies — that had not ratified it.

However, leading officials recognised that this arrangement was highly unsatisfactory, and when Yvo de Boer, a Dutch diplomat, was appointed executive secretary of the UNFCCC in 2006, he sought the support of the UN secretary-general to try a different approach, which would look beyond the Kyoto protocol and try to bring the USA on board. This resulted in a dramatic COP13 in 2007 at Bali, where negotiations dragged on far after the official deadline of 6pm on Friday had passed, and long into Saturday afternoon, as the USA delegation refused to agree to a "Bali roadmap". Finally, after a dramatic intervention on the plenary floor by Kevin Conrad of Papua New Guinea, calling on the USA to "lead, or get out of the way", the USA delegation led by Paula Dobriansky agreed to join the consensus and the Bali roadmap — which set out a pathway for a unified approach to emissions reduction beyond the Kyoto protocol — was passed. The Bali roadmap led to the Copenhagen climate summit of 2009, and an attempt at a new model of global climate action.

COP15 Copenhagen 2009: The Failure

In 2007, the Nobel Peace Prize was shared, in equal parts, between the IPCC and Al Gore for their efforts to build up and disseminate greater knowledge about man-made climate change, and to lay the foundations for the measures that are needed to counteract such change. The world had huge expectations for COP15 (Copenhagen) to deliver in 2009, despite coming just a year after the global financial crash. New quantitative commitments were expected to ensure a post-2012 agreement and seamless transition beyond the Kyoto Protocol. Barack Obama had just become President of the USA, raising hopes of a more constructive approach[1]. The EU had prepared an unconditional 20% reduction of emissions by 2020 on a 1990 baseline and a conditional target rising to 30% if other developed countries adopted binding targets. Many other developed countries also had something to offer: Norway was willing to reduce emissions by 40% and Japan by 25% from a 1990 baseline; even the USA offered a 17% reduction on a 2005 baseline, which was an equivalent drop of 4% compared with 1990. The scene was set, but the Copenhagen conference went horribly wrong: 'hopenhagen' became 'nopenhagen'. First, the Danish government underestimated the interest in the conference and provided a venue that was too small; in the second week, when all the high-powered national ministers and their entourages arrived, there was insufficient room, meaning many NGOs and observers were denied access to the negotiations. Second, it was clear that the negotiators were unprepared as there was not even a draft agreement text provided to the ministers when they arrived. This led to the leaking of 'The Danish Text', subtitled 'The Copenhagen Agreement', and the proposed measures to keep average global temperature rise to 2°C above pre-industrial levels. An argument then ensued between developed and developing nations because a brand-new text simply appeared in the middle of the conference. Developing countries accused developed ones of working opaquely and making an agreement that suited their own interests without seeking consent from developing nations. Lumumba Stanislaus Di-Aping, chairman of the G77, said, 'it's an incredibly imbalanced text intended to subvert, absolutely and completely, two years of negotiations. It does not recognize the proposals and the voice of developing countries'.

The final blow to forging a global agreement on binding targets came from the USA. Barack Obama, arriving only two days before the end of the conference, convened a meeting with the BASIC (Brazil, South Africa, India, and China) countries only, which excluded other UN nations, and created the Copenhagen Accord. This recognised the scientific case for keeping temperature rises below 2°C, but did not contain a target baseline, nor commitments for reducing emissions to achieve it. Earlier proposals that would have aimed to limit average temperature rise to 1.5°C and cut carbon dioxide emissions by 80% by 2050 were dropped. But all mention of a 1.5°C was continually blocked by China. The resulting agreement was non-binding and countries could provide their own voluntary targets. It was also made clear that any country that signed up to the Copenhagen Accord was also effectively stepping out of the Kyoto Protocol, hence the USA was able to move away from the binding targets of Kyoto Protocol, which should have been enforced until 2012. A weak voluntary commitment approach was adopted. The Bolivian delegation summed up the way the Copenhagen Accord was reached: 'anti-democratic, anti-transparent and unacceptable'. It was also not clear what legal status the Copenhagen Accord had because it was only 'noted' by the parties, not adopted, as only 122, subsequently rising to 139 countries, agreed to it.

Trust in the UNFCCC negotiations had another setback when, in January 2014, it was revealed that the USA Government negotiators had information during the conference obtained by eavesdropping on meetings of other conference delegations. Documents leaked by Edward Snowden suggested the USA National Security Agency (NSA) had monitored communications between countries before and during the conference. The leaked documents revealed that the NSA provided the USA delegates with advance details of the Danish plan to 'rescue' the talks should they flounder, and laid bare China's efforts to coordinate its position with India before the conference.

Although Copenhagen fell short of its aims, the accords produced — which were formally adopted under the UN process at COP16 in Cancun the following year — did mark the first time that developed and developing countries (non-Annex 1 countries) jointly accepted commitments to take action on GHG emissions. In the case of developed countries, those were emissions cuts; for developing countries, curbs on the future growth of their emissions.

COP16 Cancun and COP17 Durban: Progress is Made

At COP16 in Cancun, the UN tried to repair some of the damage of COP15, adopting most of the commitments made in Copenhagen as COP decisions, which formalised them under the UN process. At COP17 in Durban, most countries were expecting a somnolent affair, but Connie Hedegaard, the EU climate commissioner, who in her previous role as Danish environment minister had been president of COP15, wanted a different approach. Recognising that the Obama presidency offered a limited opportunity to sign a legally binding agreement, she came to Durban with a plan for a fresh roadmap, to what would eventually become the Paris agreement. Durban marked a bifurcation among developing countries, which had previously tended to follow the lead provided by China. Many, particularly among the poorest, were increasingly realising that their interests were not aligned, as China's policies were allowing vast increases in GHG emissions, of which they would bear the brunt. The EU managed to gather a broad coalition of more than 100 of such developing countries and had the tacit approval of developed counterparts such as the USA, Japan and Canada, behind its push for a new legally binding agreement. By the scheduled end of the conference, it was clear that two big countries were holding out: China and India. The talks dragged on far beyond their deadline, through Saturday and into the early hours of Sunday, with many countries predicting that the EU would cave, and China and India would succeed in stymying progress. But in the end, no longer able to call on allies among the least developed countries, China and India gave way, and the process towards what became the Paris agreement was begun.

COP21 Paris 2015: The Paris Agreement

The failure of COP15 in Copenhagen and its voluntary commitments cast a long shadow over subsequent COP meetings, compounded by the Wikileaks revelation that the USA had decreased funding to Bolivia and Ecuador was because of their opposition to the Copenhagen Accord. It would take over five years for the negotiations to recover from the difficulties created at Copenhagen. At COP16 in Cancun and COP17 in Durban, the UNFCCC negotiations were slowly rejuvenated with the aim of agreeing to a new round of negotiations and the underlying principles. There were some important political breakthroughs, including the agreement on the 2°C target, establishment of the Green Climate Fund, the setting of the $100bn a year funding target, and the notion of voluntary emission pledges for all — the basic backbone of the Paris agreement. In parallel, significant progress was made in REDD (Reduced Emissions from Deforestation and Forest Degradation), including safeguards for local people. It was not, however, until COP18 in Doha in December 2012, that a second commitment period for the Kyoto Protocol lasting eight years was agreed, to commence from January 2013. This ensured that all Kyoto mechanisms and accounting rules would remain intact, and that parties could review their commitments with a view to increasing ambition. All this laid the foundations for the possibility of a future global climate agreement, which was finally agreed at COP21 in Paris in 2015.

The climate negotiations in Paris 2015 were a huge success primarily because the French hosts understood the grand game of international negotiation and used every trick available in the diplomatic playbook to get the countries working together to achieve an agreement signed by all. The Paris Agreement set a goal to hold temperatures to "well below 2°C above pre-industrial levels" and "pursue efforts to limit the temperature increase to 1.5°C above pre-industrial levels". Paris was a high-stakes game of geopolitical poker. Surprisingly, the least powerful countries did much better than expected. The climate talks were subjected to a series of shifting alliances going beyond the usual dichotomy between income-rich global North and income-poor global South countries. Central to this was, firstly, USA-Chinese diplomacy — both agreed to limit emissions in 2014 under the U.S.A.-China Joint Presidential Statement on Climate Change. Secondly, a new grouping of countries under the banner of The Climate Vulnerable Forum, including Small Island Developing States, pushed the 1.5°C target higher up the political agenda, so much so that it is mentioned in the key aims of the agreement. Political and civil society support generated from the success of the Paris Agreement created the preconditions for the IPCC to land its special report on Global Warming of 1.5°C to an attentive global community. Highlighting not just the stark differences between the impact of a 1.5°C and 2°C world, the report also modelled pathways of how a 1.5°C world could be achieved: (1) a 45% decline in CO2 emissions from 2010 levels by 2030; (2) net zero CO2 emissions by 2050; and (3) thereafter for the rest of the century. The quicker the world could get to net zero, the less CO2 that would need to be sequestered from the atmosphere between 2050 and 2100.

Crucially, the Paris Agreement was intended to be the start of a process of ratcheting up mitigation ambition[6]. Collectively, initial pledges were necessary but far from sufficient. Even as late as 2019, assuming all national mitigation pledges were fulfilled, the world would still warm by about 3°C or more.

COP26 Glasgow 2021: The Glasgow Climate Pact

This new wave of public global environmental concern meant Copenhagen-esque expectations surrounded COP26 in Glasgow, co-hosted by the UK and Italy, and originally scheduled for 2020. But due to COVID-19 enforced restrictions and lockdowns, and the severe impacts on both Italy and Britain, this pivotal meeting was postponed until November 2021. COP26 in Glasgow was a different type of negotiation from that of COP21 in Paris, which set the foundations for a new global climate agenda and ambition architecture. Instead, COP26 was critical because it was the third meeting of the parties to the 2015 Paris Agreement (CMA3) and the first-time countries had agreed to raise the ambition of their carbon-cutting pledges, as outlined in the Paris Agreement[7]. Importantly, the focus of COP26 was on 'net zero' carbon emission targets and the pathways to them for each country.

Despite 2020 and 2021 being dominated by COVID-19, the geopolitical landscape around climate change had shifted seismically before COP26. First, in June 2019, the UK parliament amended the Climate Change Act (2008) to require the government to reduce the UK's net emissions of GHGs by 100% relative to 1990 levels by 2050 (net zero). Second, the European Commission agreed the EU would reduce its GHG emissions by at least 55% by 2030 from 1990 levels, instead of the 40% cut agreed six years prior. Third, in September 2020, China's President Xi Jinping announced via video-link to the UN General Assembly that his country would aim to peak CO2 emissions before 2030, followed by a long-term target to become carbon neutral by 2060. China, the world's largest annual emitter of CO2 and accountable for around 28% of global GHG emissions, had, up until then, not committed to a long-term emissions goal. Under the Paris Agreement, China had merely pledged to cut the carbon intensity of its economy by 60–65% against a 2005 baseline. This announcement followed long and detailed discussions between China and the EU on climate change.

COP26 also marked the re-engagement with the USA, the second largest emitter of global GHG emission. President Trump had begun the process of removing the USA from the Paris Agreement in 2017, but in accordance with Article 28 of the agreement, a country can only give notice of withdrawal 'after three years from the date on which [the] Agreement has entered into force'. So, the earliest possible effective withdrawal date by the USA was November 4th 2020, one day after the 2020 U.S.A. presidential election. Newly elected President Biden immediately cancelled the withdrawal and became a strong advocate of collective international action to deal with climate change.

In many ways COP26 was a small step forward, with little or no back sliding. The Glasgow Climate Pact agreed at the conference includes a strong statement on the necessity of achieving the 1.5°C target — including a renewed call for a 45% reduction in CO2 emissions by 2030. The Pact also includes a call to 'phase down' coal and remove inefficient fossil fuel subsidies. While the western press made a lot of fuss over the late switch in language from 'phase out' to 'phase down' coal (due to pressure from India and China), they missed the fundamental shift in COP terminology: it represented the first time that fossil fuels had ever been mentioned in an international climate agreement. There was also a call in the Pact for new carbon-cutting pledges (National Determined Contributions, NDCs) for COP27 to boost global ambition in the near-term. As stated in the Paris Agreement, NDCs are only required every 5 years, so the call for new NDCs by COP27 suggested updates could become annual or biennial to try and accelerate decarbonization. In Glasgow, the much-maligned and long-awaited Article 6 rulebook was finally finished and signed. This complicated agreement stipulates how countries should monitor and account for their carbon emissions and reductions. It, in effect, allows countries to sell their emission reductions to others, while at the same time preventing double accounting, providing a robust framework within which countries and companies can trade carbon. While loopholes still exist, there was an agreement which everyone could work with finally.

Still, COP26 had notable failures. First, the developed nations failed to honour their 2009 commitment to provide $100 billion per year to help the least developed countries transition from fossil fuels to clean energy. Over a decade after its ratification in 2010, this promise has yet to be fulfilled. Second, COP26 failed to create the Glasgow Loss and Damage facility, whereby least developed countries could claim compensation from historically high GHG-emitting countries for damage caused by climate change — the agreement was blocked by the USA and the EU on the conference's last day. Third, countries' carbon-cutting NDCs are not legally linked to the Paris Agreement 1.5°C pathway, and despite shifts in some countries' positions, even if all pledges are fulfilled, the world is still looking at heating of between 2.4–2.8°C by the end of the century.

Beyond the Glasgow Climate Pact, other international agreements were drawn up on, for example, reducing deforestation, phasing out coal and removing coal finance. An agreement to cut methane emissions by

30% in the next 10 years was also secured. There was, in addition, an announcement that the USA and China would collaborate more closely on climate change throughout the rest of this decade — that is important as, together, they represent 43% of the world's GHG emissions — but no details have been forthcoming of this collaboration.

Disruptive Influence of the Russian Invasion of Ukraine

Geopolitics have altered rapidly in the last few years, with one of key shocks being the Russian invasion of Ukraine. In terms of climate change, a mixed response is emerging: the EU is moving away from Russian gas as quickly as possible, having pledged to double the instillation of renewable energy this decade. Meanwhile, other countries such as the USA started to ramp up production to be self-sufficiency in fossil fuels, while China, India, and Turkey, wishing to boost their domestic economies, leaped at the opportunity to buy cheap Russian oil caused by western sanctions on Russian exports. Imports of Russian oil rose by over 50%, displacing Saudi Arabia as China's biggest oil provider. The Russian invasion meant that oil and gas-producing nations became more influential at COP27 and COP28, undermining the negotiations. World leaders preoccupied with spiraling energy prices and the escalating cost of living were reluctant to act boldly on fossil fuels. This was reflected in the watered-down text in which the Egyptians slipped in a provision to boost "low-emission and renewable energy", which is a nod to natural gas (cleaner than oil and coal but still a fossil fuel).

In the longer term, the invasion of Ukraine has put energy security back on the top of governments' agendas. For countries with no or little access to domestic fossil fuel reserves, renewables are set to become very attractive — they are already much cheaper to build and maintain than coal-fired power stations.

COP27 Egypt: A Miss Opportunity

In 2022, two major IPCC reports were published — IPCC Working Group II's 'Impacts, Adaptation and Vulnerability[8]' and Working Group III's

'Mitigation of Climate Change[9]'. Both painted an exceptionally stark future if GHG emissions are not reduced rapidly. Importantly, the reports also demonstrated that almost all the solutions required to mitigate and adapt to climate change already exist — they just need to be scaled up. The possible futures these reports illustrated provided the backdrop to COP27 in Sharm El Sheik, Egypt, billed as the 'African COP'[10]. The African continent is already experiencing major climate change, including average temperature rises higher than many other parts of the planet, and the continent will suffer some of climate change's most severe future impacts. At COP26, countries were requested to submit revised carbon-cutting NDCs by COP27 to increase global ambition on mitigating emissions and close the ambition gap necessary to meet the temperature targets of the Paris Agreement. But the COP27 agreement failed to go beyond the 2021 Glasgow Climate Pact's promise to "phase down unabated coal power", despite the Indian proposal that all fossil fuels should be phased out. The text also announced no new targets or commitments, threatening the goal of limiting global temperature rise to 1.5°C, established seven years ago in the Paris agreement. Instead, there was a request for new country pledges, or nationally determined contributions (NDCs), for COP28 — another year's delay.

Developing countries entered COP27 with hopes for progress on three fronts: climate finance and the delivery of US$100 billion a year as promised in 2009, global decarbonisation, and recognition of the responsibility of developed countries to pay for loss and damage. Only one of these was achieved to any degree.

COP27: Loss and Damage

Loss and damage refer to the most devastating ravages of extreme weather, so severe that no amount of adaptation can help to avoid them[2]. Examples include the devastating floods that hit Pakistan and the severe droughts in Africa in 2022, the same year as COP27. Recovery from such devastation can take years, if it is ever achieved, and the infrastructure of developing countries, services such as health and education, and their chances of improving the lives of their people, can suffer permanent damage.

The world's poorest countries, which have done least to cause the climate crisis, are most at risk from loss and damage. In the past, some experts characterised loss and damage as a form of compensation or reparations for poor countries, from the rich. However, this was unacceptable to developed and large developing countries, which refused to sign up to the Loss and Damage facility because it may potentially leave them liable for unlimited future costs despite the fact that this was specifically addressed in the Paris Agreement which states that Loss and Damage does not involve or provide a basis for any liability or compensation.

The discussion has moved on to loss and damage as a form of rescue and rehabilitation for the countries suffering the most, differing from climate finance in that it does not apply to emissions cuts, and addresses broader social and development issues as well as the immediate impacts of extreme weather. For years, little progress has been made on loss and damage discussions, but at COP26, developed countries signaled they would discuss new finance mechanisms for loss and damage. Though COP27 will be viewed as a failure, in the early hours of Sunday morning, well past the Friday deadline, member states agreed to establish such a fund — a win for developing countries. Who will pay and how this financial assistance will be delivered to help countries like Pakistan recover from climate disasters remains to be negotiated.

Why COP27 Failed to Deliver

First, the timing of COP27 was unfortunate[10]. Week one occurred during the USA midterm elections when much of the world's media was scrutinising its finely balanced outcome. Week two coincided with the G20 summit in Bali, which further diverted attention, and meant that many world leaders did not attend COP27. To make matters worse, negotiations stretched into the weekend, just when attention turned to the World Cup and associated controversies in Qatar. This was very different from COP26 when the world remained engaged throughout the summit. The only protests allowed were those sanctioned by the Egyptian security forces within the venue. With media attention already restricted, the limited but important civil society presence at COP27 struggled to keep pressure on the hosts.

Second, there was a lack of leadership. International diplomacy is difficult and takes a huge amount of time, effort and skill. The reason why 2021's COP26 in Glasgow yielded agreements on deforestation, methane emissions and other issues was partly because the UK and Italian hosts worked hard to build consensus during the extra year provided by the pandemic. Egypt's presidency of COP27 underestimated this task. When the negotiations carried over to the wee hours of Sunday morning, Egyptian COP27 president, Sameh Shoukry, said: "It is really up to the parties [countries] to find consensus." This is in stark contrast to COP26, where the president of the conference, Alok Sharma, fought to the bitter end to secure an agreement. Negotiations were only ramped up in the last 48 hours to get an agreement on loss and damage, and even then, some of the larger emitters (China and India) have refused to contribute to the fund.

Third, there is a lack of trust between the developing and developed countries. This is primarily because the US$100 billion promised per year has yet to fully materialise. This is a relatively small amount of money when you consider Qatar has reportedly spent $220 billion alone to host the 2022 World Cup. Money to support climate change adaptation has also not been forthcoming. The money is there; the issue is the will to allocate it where it is really needed. And the biggest sticking point was over loss and damage. At COP26, the USA, EU and UK, with support from China, blocked the setting up of the Glasgow loss and damage facility, as they did not want to be liable for the effects of climate change. In Egypt, a statement was released at the last-minute saying that such a loss and damage fund would be set up after all. It was a step in the right direction and was celebrated by developing nations. But there was no agreement about how large the funding stream would be, who pays, and critically, who controls and manages these funds. Currently, only 10% of climate finance reaches local communities, and the new facility will need to address this disconnect. Countries like China and India pushed back on contributing to those funds. India resisted the inclusion of terms such as "current high emitters" in the text as it expects historically high emitters to contribute to the funds. This may have also been the case for China 30 years ago. But China's historic emissions are nearly as high as the

EU's now, so its leaders point to per capita emissions and has restated its status as a developing country.

COP28 Dubai: Transition Away from Fossil Fuels

COP28, was always going to be controversial[11]. It was held in a state whose economy is heavily dependent on oil and gas — the United Arab Emirates. The summit president, Sultan Al Jaber, remains head of the Abu Dhabi National Oil Company and recently declared the UAE would double oil and gas output this decade. This was hardly the strong leadership needed to move away from fossil fuels that many had hoped for. More than 100,000 delegates were registered (twice as many as any previous COP) of which more than 2,000 were official representatives of fossil fuel companies.

Perhaps surprisingly, this was the first COP to officially acknowledge that fossil fuels are the root cause of climate change. And it is worth remembering that fossil fuels were only first mentioned in an international climate agreement in 2021 at COP26 in Glasgow. Yet it still lacked ambition. Most countries wanted a strong statement on phasing out or at the very least, phase down (reducing) fossil fuels. Instead, countries are forced to agree on a statement saying we must "transition away from fossil fuels in energy systems, in a just, orderly and equitable manner, accelerating action in this critical decade, so as to achieve net zero by 2050 in keeping with the science".

The "global stocktake" was finalised at COP28[11]. This was the first time the global climate regime took stock of how the international community has collectively reduced its GHG emissions since the Paris Agreement in 2015. The biggest takeaway from the stocktake told us what we already knew — the world is way behind and the 1.5°C warming limit agreed in Paris is at risk. Loss and damage fund that was agreed at COP27 in 2022 had an injection of US$700 million being pledged. While this was welcome news, it was still a drop in the ocean compared with the US$400 billion actually needed. COP28 did not sort out how the fund will work, what the major funding streams will be or whether the allocation of finance will be community-driven and corruption free. And despite opposition, it had been agreed that the World Bank will administer the fund for a negotiated fee of 24% — meaning one in four dollars pledged will never make it to the countries in need.

COP29 Baku: The Finance COP

Billed as the "finance Cop", the COP29 was held in Baku, Azerbaijan, and was expected to provide the money to enable the transition away from fossil fuels announced at COP28[12]. Negotiations were focused on increasing the US$100 billion (£79.4 billion) promised per year by developed countries to help the least developed countries build up their renewable energy capacity. Many demanded a ten-fold increase, as at least US$1 trillion is needed per year. Discussions also centered on who should finance the loss and damage fund agreed at COP27 that would compensate developing countries for the impacts of climate change.

The conference was a mixed bag. First, the money never really showed up[12]. There is an urgent need to invest trillions of dollars in climate mitigation and adaptation — yet even though this was the finance COP, very little money was put on the table. The developing countries asked for $1.3 trillion from developed countries by 2035, with added voluntary contributions from better-off countries still classed as developing, such as China. What they got in the early hours of Sunday, after the summit had overrun by 33 hours, was a much lower pledge of US$300 billion by 2035 provided by governments and the private sector. COP29 did get the loss and damage fund up and running, but it remains woefully underfunded, like many of these financial mechanisms, and no agreement was reached on who pays in and who can claim.

Overall COP29 can be defined as — China steps up, USA says goodbye, and India stays at home. First, it is clear that China means business when it comes to carbon neutrality and investment in green technologies. That much was evident from strong engagement at COP29 from its ministries, private sector and experts. However, it wants to stick to definitions of "developing nations" used when the UN's climate framework was devised back in 1992. Under that definition, China — now the world's second largest economy and large emitter of carbon

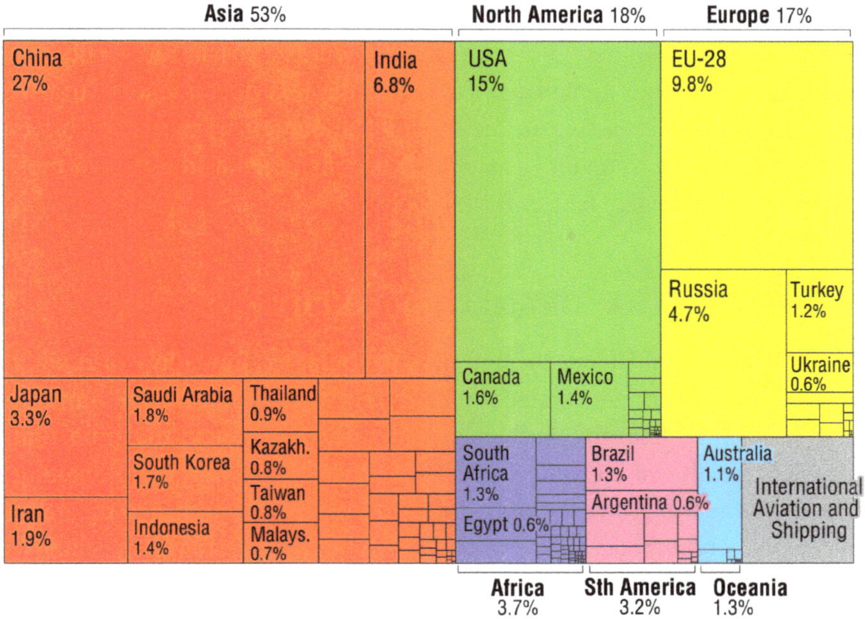

Figure 53. Country Carbon Dioxide Emissions by Percentage

dioxide (Figure 53) — is still developing and need only make voluntary contributions to climate finance.

India, on the other hand, dismissed COP29 as a "technical COP", with the senior leadership staying at home in part due to elections in the state of Maharashtra. India's delegation said that the US$1.3 trillion investment should come from developed nations, which does not include India. And the USA said goodbye, again, due to the election of Donald Trump as the USA's President again. Many have suggested the lack of financial ambition is in part because the USA will not be contributing to the funds during a Trump presidency. And of course, Saudi Arabia tried to wreck the party. COP28 climate summit agreed on a historic resolution to "transition away from fossil fuels". Yet from the beginning of COP29, the Saudis tried to ensure that it was not restated, much to the annoyance of the UK, the EU and many other countries — including even fellow petrostate (and COP28 host), the UAE. At the final hour, the "transition away from fossil fuels" resolution was included in the agreement but only by referring to it by paragraph and document number, so the words "fossil fuels" do not appear in the COP29 agreement. However, the text does include "transitional fuels can play a role in facilitating the energy transition". This is largely a reference to natural gas, which is considered a transitional fuel. Sadly, another win for the fossil fuel lobby.

COP30 Brazil: A Missed Opportunity

COP30 was held in Belém located in Amazonia, to show the world what the frontlines of climate change look like. This was the first COP Summit since COP26 in Glasgow that was held in a democracy, meaning that protests were allowed. Over 5,000 Indigenous people were there, making their voices heard. However, only 360 of them secured passes to the main conference, compared to 1,600 delegates linked to the fossil fuel industry. Inside the negotiating rooms, it was business as usual, with Indigenous groups remaining as observers, unable to vote or attend closed-door meetings. But the highly visible protests did help obtain recognition of four new Indigenous territories in Brazil. It shows that when civil society has a voice, it can secure wins, even outside of the main climate negotiations.

The absence of the USA at COP30 did create a vacuum but also opportunity. In Donald Trump's first turn as president, the USA sent at least a skeletal group of negotiators to the COP meeting. This time, in a historic first, America did not send an official delegation at all. This is because Trump recently described climate change as "the greatest con job ever perpetrated on the world". The USA is now slowing renewables while expanding oil and gas. This is allowing other oil producing countries like Saudi Arabia to ignore their own climate pledges and to try undermining others. China stepped into this void, and at COP30, was one of the loudest voices in the room. As the world's largest supplier of green technology, China used COP30 to promote its solar, wind and electric vehicle industries, and engaged with countries looking to invest.

COP30 was also referred to as the Implementation COP — with many side deals achieved. The Belém pledge, backed by countries including Japan, India and Brazil, committed signatories to quadruple sustainable fuels production and use by 2035. Brazil also launched a major trust fund for forests, with around US$6 billion (£4.6 billion) already pledged for communities working to protect rainforests. The EU followed by pledging new funds for the Congo Basin, the world's second largest rainforest.

COP30 also started well with the Brazilians putting forward the Global Mutirão text, a proposed roadmap to "transition away" from fossil fuels. More than 80 countries signed it, from EU members to climate-vulnerable Pacific Island states. But opposition from Saudi Arabia, India and other major fossil fuel producers watered it down. The final outcome of COP30 — the Belém package — was weak and will get us nowhere near the 1.5°C Paris agreement limit. Most striking is the absence of the words "fossil fuels" from the final text even though they were central to the Glasgow climate pact (2021) and the UAE consensus (2023) — and of course they represent the main cause of climate change. COP30 was a divisive climate summit. The gulf between oil-producing countries (particularly from the Middle East) and the rest of the world has never been wider.

Is the UNFCCC Process Flawed?

Various flaws have been pointed out in the approach of the UNFCCC. Here are some of the major ones[13].

Not going far enough. The first perceived flaw in the UNFCCC procedure is that despite 35 years of negotiations, it has failed to deliver any lasting agreement. As mentioned above, with the current Paris Agreement, even if all the pledges are honoured, will still cause global warming of 2.8°C and the associated impacts described in Table 4.

No enforcement. The fundamental problem with international agreements and treaties is that there are no real means of enforcement. This was one of the arguments that the USA used when proposing the Copenhagen Accord, suggesting that even binding targets must in effect be voluntary as countries decide whether or not to comply. This is why policies and laws are required at a regional level, as in the EU, and at a national level, such as the UK's Climate Change Act. The only way to uphold international treaties is through regional and national laws. This multi-level governance is also required to stop any countries from gaming of particular systems.

Green colonialism. Many social and political scientists have raised philosophical and ethical doubts about climate negotiations as a whole. The main concern is that they reflect a version of colonialism, since rich developed countries are seen to be dictating to poorer countries how and when they should develop. For many years, countries such as India and China resisted calls to cut their emissions, stating that it would damage their development and attempts to alleviate poverty. It is only since the Paris agreement of 2015 that most countries have acknowledged the need for the global net zero target. Early on, there was support for measures such the Clean Development Mechanism (CDM), which allows developed countries to pay for emission reductions in a developing country, and it counts towards their national reduction target. These measures also provide a development dividend, moving money from the rich to the poorer countries. However, 80% of the CDM project credits were allocated to China, Brazil, India, and Korea, the richest developing countries, so funding did not reach the world's poorest. Also, 60% of the CDM carbon credits have been purchased by the UK and the Netherlands, resulting in a very skewed financial exchange.

Nation versus sector approach. The UNFCCC approach has another problem, which is embedded in the concept of the nation-state and is a major issue in a global capitalist world with supposedly free trade. For example, if the USA through the Paris Agreement wanted to reduce carbon emissions from heavy industry, it could impose a carbon tax on steel and concrete production. If other countries in the world do not have this restriction, their products become cheaper, even including the cost of transportation by ship, air, or road to the USA, all of which would lead to the emission of more carbon dioxide overall. Therefore, global economics can undermine any national attempts to do the right thing in reducing their emissions. An alternative approach would be for global agreements to be made at a sector level. For example, there could be a global agreement on how much carbon can be emitted per ton of steel or concrete produced.

All countries could then agree only to buy steel or concrete produced in this low emission way, which would make for a fairer trading scheme, with countries not losing out as a result of changes within their industries to lower GHG emissions. Critics point out that this would, of course, lead to a rise in the price of both steel and concrete as manufacturers would have to pay for the production process. It may even lead to shortages as very little of today's concrete and steel is made in a way to reduce GHG emissions.

Summary

In the last 30 years, the amount of human-emitted CO_2 has doubled, representing a collective failure of the world's leaders to rise to Margaret Thatcher's 1989 global call to action. However, the climate negotiations and the resultant changes in the global economy mean that predicted levels of global warming above 4°C this century are now highly unlikely. If all the country pledges that were made at Glasgow COP26 and reinforced at COP30 are fulfilled, then global warming would be kept to between 2.4°C and 2.8°C. This is nowhere near the 1.5°C target agreed at the COP21 in Paris, but we have to acknowledge that without these international negotiations, the ambition of getting to 'net zero' global emissions by the middle of this century would not exist. At this moment in time, 84% of the world GDP and over 75% of global GHG emissions are under a net zero emissions pledge. These figures include individual USA states that have pledge to go net zero but not the USA as a whole as the current administration is backtracking on previous promises. We now need to ensure that each country makes good on their pledges and accelerates their progress, and if possible, bring the USA back into the climate negotiations. The next chapter discusses all the different ways countries can mitigate their emissions and how it can increase prosperity, security as well as human health and wellbeing.

Chapter 11

Solutions: Mitigation

Introduction

There are three types of solutions to climate change. The first is mitigation, which in its simplest terms, is reducing our carbon footprint and thus reversing the trend of ever-increasing GHG emissions. The second is adaptation, which is providing protection for the population from the impacts of climate change. Third is geoengineering, which involves large-scale extraction of carbon dioxide from the atmosphere or modification of the global climate. This chapter focuses on mitigation and pathways to net zero emission by the middle of this century.

Mitigation

The requirement to cut global carbon emissions by half by 2030 and hit net zero by 2050 is extremely challenging, and will mean using every available solution as soon as possible[1]. There is some good news. GDP growth and carbon emissions have over the past decade become completely decoupled, with a large rise in world GDP compared with a much smaller rise in carbon emissions. What we need to do now is to invert the relationship so that as GDP rises, carbon emissions drop. In 2020, the International Energy Agency (IEA) and International Monetary Fund (IMF) published a report recommending massive investment into clean energy, which would create millions of new jobs[2]. It was realised during the Covid-19 pandemic that energy generation and use was the key to dropping carbon emissions. The report outlines various plans for mass home renovations, fossil-fuel subsidy reforms, expansion of renewable energy, and power grids. Several of these are discussed below.

Alternative, Renewable, or Clean Energy

Fossil fuels are an amazing discovery, and they have allowed the world to develop at a faster rate than it has at any other time in history. The high standard of living in the developed world is based on cheap and relatively safe fossil fuels. Unfortunately, burning fossil fuels has the unintended consequence of changing global climate. Therefore, in the 21st century, we need to switch from fossil fuel energy to low-carbon or carbon-neutral energy. These include solar, bio, wind, hydro, wave, and tidal energy[1].

Solar. The Earth receives on average 343 W/m^2 from the Sun, and yet as a whole, the planet only receives two-billionth of all the energy put out by the Sun. The Sun is the ultimate source of energy, which plants have been utilising for billions of years. At the moment, we can convert solar energy directly to heat or electricity, or we can capture the energy through photosynthesis by growing biofuels[4]. The simplest approach is through solar heating. On a small scale, houses and other buildings in sunny countries can have solar heating panels on the roof, which heats up water, so people can have carbon-free hot showers and baths. On a large scale, parabolic mirrors are used to focus the solar energy to generate hot liquid (water or oil) to drive turbines to create electricity. The best places to situate solar heat plants are in low-latitude deserts, which have very few cloudy days per year. Solar heat plants have been built in California since the 1980s and are now being built in many other countries. Solar photovoltaic panels convert sunlight directly into electricity. The individual rays of the Sun hit the solar panel and dislodge electrons inside it, creating an electrical current. The main advantage of solar panels is that you can place them where the energy is needed and avoid the complicated infrastructure normally required to move electricity around. Over the past decade, there has been a massive increase in their efficiency, the best commercially available solar panels being about 23% efficient, which is significantly more than photosynthesis at about 1%. Solar panel efficiency goes up during winters, as they work better in cold temperatures, but of course, produce less electricity due to shorter days and less intense sunlight. There has also been a significant drop in price due to huge investment in the technology.

Biofuels. These are the product of solar energy converted into plant biomass via photosynthesis, which can then be used to produce liquid or solid fuels[1]. The global economy is based on the use of liquid fossil fuels, particularly in the transport sector. So, in the short term, fuels derived from plants could be an intermediate low carbon way of powering cars, ships, and airplanes. Ultimately, electric cars are the future, because the required electricity can be produced carbon-neutrally. However, this energy source is not an option for airplanes. Traditional air fuel 'kerosene' combines relatively low weight with high energy output. Research is being carried out to see whether a biofuel can be produced that is light enough and powerful enough to replace kerosene. Many power stations around the world have been converted so that they can burn wood pellets instead of coal or natural gas to create steam to turn the turbines to make electricity. However, the scientific evidence is clear that wood pellets are not sustainable, and some wood pellets from the USA even have the same carbon footprint as natural gas.

Wind. Wind turbines are an efficient means of generating electricity, if they are large and preferably located out at sea. Ideally, the larger the turbines, the better the effectiveness and efficiency. The largest onshore windfarm is in Jiuquan, China, with a planned installed capacity of 20 GW. The wind farm, also called the Gansu Wind Farm, will consist of 7,000 wind turbines, which will be installed in the remote desert areas in Gansu Province which is near the Gobi Desert. The largest operation offshore wind farm is the Hornsea Wind Farm Zone, located off the coast of Yorkshire, England. Specifically, Hornsea 2 is the largest operational offshore wind farm, boasting a capacity of 1.3 GW and is capable of powering over 1.4 million UK homes, reducing harmful carbon dioxide emissions by nearly 3 million tonnes per year. In 2026 the UK Government held an auction for offshore wind and another 8.4 GW of wind power contracts were awarded. However, there are some problems with wind turbines. Firstly, they do not supply a constant source of electricity; if the wind does not blow or it blows too hard, no electricity is generated. Secondly, people do not like them, finding them ugly, noisy, and a worry in terms of the effects they may have on local natural habitats. Thankfully, all these problems are easy to overcome by situating wind farms in remote locations, out at sea, and away from areas of special scientific or natural interest. Recent research has shown little or no effect on local wildlife even when wind turbines are situated close to land. One study suggests that wind in principle could globally generate over 123 petawatt-hours[5], which is over five times the current global electricity requirement. Though subsequent criticisms suggest this is an over-estimation, it did highlight that wind power could potentially provide all our current electricity needs.

Wave and tidal. Wave and tidal power could also be an important source of energy in the future[6]. The concept is simple: to convert the continuous movement of the ocean in the form of waves into electricity. However, this is easier said than done, and experts in the field suggest that wave power technology is 20 years behind solar panel technology. Tidal power has one key advantage over solar and wind — it is constant. In any country, for energy supply to be maintained at a constant level, there has to be at least 20% production guaranteed, known as the baseline requirement. With the switch to alternative energy, new sources of this consistent baseline need to be developed.

Hydro-energy. Hydroelectric power is globally an important source of energy; it currently supplies 15% of the world's electricity[6]. The majority of the electricity comes from large dam projects. These projects can present major ethical problems as large areas of land must be flooded above the dam, causing mass relocation of people and destruction of the local environment. A dam also slows water flowing down a river and prevents nutrient-rich silt from being deposited lower down. If the river crosses national boundaries, there are potential issues over the rights to water and silt. For example, one of the reasons why Bangladesh is sinking is the lack of silts due to dams on the major rivers in India. There is also debate on how much GHG emissions hydroelectric plants save, because while the production of electricity does not cause any carbon emissions, the rotting vegetation in the area flooded behind the dam does give off significant amounts of methane. Hydropower is also an important way to store energy.

As renewable energy can be variable and intermittent, it is important to store energy when it is produced so it can be used when needed.

Geothermal. Below our feet, deep within the Earth, is hot molten rock. In some locations, for example in Iceland and Kenya, this hot rock comes very close to the Earth's surface and can be used to heat water to make steam. This is an excellent carbon-free source of energy, because part of the electricity generated from the steam is used to pump the water down to the hot rocks[6]. Unfortunately, it is currently limited by geography, and nothing can be done until deep drilling technology is developed to expand the areas that can access geothermal energy. There is, however, another way the warmth of the Earth can be used. All new buildings can have boreholes below them with ground-sourced heat pumps. This involves drilling deep vertical holes (often 40–200m) to install small heat exchange pipes usually about 15 cm in diameter, providing a compact, efficient way to extract ground heat for heating and hot water. Cold water is pumped down into these boreholes and the ground warms the water up, cutting the cost of providing hot water to building. The beauty of this method is that it can be used almost everywhere in the world and for properties that have limited space.

Nuclear fission. Energy is generated when heavy atoms such as uranium are split; the process is known as nuclear fission. The process has a very low direct carbon signature, but a significant amount of carbon is generated when mining the uranium, building the nuclear power station, decommissioning the power station, and safely storing and disposing of nuclear waste. At the moment, 5% of global energy is generated by nuclear power[6]. The new generation of nuclear power stations are extremely efficient, achieving nearly 90% of the theoretically possible energy production. The main disadvantages of nuclear power are the generation of high-level radioactive waste and concerns about safety, though improvements in efficiency has reduce waste and the new generations of nuclear reactors have state-of-the-art safety precautions built in. The 1986 Chernobyl disaster and the 2011 Fukushima Daiichi nuclear disaster illustrate that nuclear plants are still not safe, being vulnerable to human error and natural hazards. The advantages of nuclear power, however, are that it is reliable and can provide the required base load in the energy mix, and it is technology that is already available and thoroughly tested.

Nuclear fusion. This process involves the generation of energy when the nuclei of two small atoms fuse together. It is the process that lights up our Sun and every other star. The idea is that deuterium, the heavy form of hydrogen found in seawater can be combined with tritium, the other heavy isotope of hydrogen, and the only waste product is the non-radioactive gas, helium. The problem, of course, is persuading those nuclei to fuse. In the Sun, fusion occurs in the core, at incredibly high temperatures and pressures. Great advances in fusion technology have been made around the world, but what is required is a huge investment to make fusion commercially viable. It is now a race between the USA, UK and China to be the first to build a fusion power station and all are aiming for 2040 at the latest.

Batteries. Batteries are included in this list because they are crucial for global renewable energy transition. Renewable energy such as wind and solar are intermittent and energy storage will be essential to stabilise national energy grids. Batteries are essential for powering electric vehicles (EVs) and help decentralise power through mini-grids, ensuring more reliable, cleaner energy supply as electrification expands. A recent study suggested that the EU only has about 25 gigawatts of total battery storage, most of which is in individual homes, with only five gigawatts of large utility-scale battery storage. To achieve the EU's net zero goals, it is estimated another 130 gigawatts of storage will be needed by 2030.

Transport

One of the greatest challenges to mitigating GHG emissions is transport[4]. At the moment, transport accounts for 14% of GHG emissions globally. In many developed countries, the carbon emissions from energy production, business, and residential sectors are all going down despite annual growth in the economy; but transport emissions, mainly from cars and aviation, are still growing. Many in the developing world aspire to the same level of car ownership and international travel as the developed world, and hence there is the potential for huge growth in transport emissions.

Electric cars, both in terms of range and performance have improved greatly over the past decade and there is general acceptance that they are the future. This acceptance has been accelerated by the 2020 pandemic, during which in many regions, road traffic almost ceased due to lockdowns, and everyone noticed the huge improvement in air quality. If there were a switch to 100% electric vehicles, there would be a 50% cut in air pollution. Air pollution unfortunately is still created by the wear of tyres and breaks, while tarmac on roads account for the other 50%. The impact of electric cars on carbon emissions can be significant, but relies on there being a guaranteed supply of low-carbon or carbon-neutral electricity. Currently, China leads the world in total number of EV uptake, with the share of EVs in its total new car sales exceeding 40% in recent years. Though a much smaller market, Norway's EV uptake is nearly 90% of new car sales in 2024 and about a third of all cars on the road by early 2025.

International shipping and aviation account for 3.2% of global GHG emissions per year. Airplanes have become an easy target for climate change campaigners as international flights are a highly visible symbol of consumption and have never been covered by an international treaty. There is a need for incentives to improve the carbon efficiency of flights, and ultimately to make them as close to carbon neutral as possible. The fundamental issue is that currently, an international treaty prohibits the taxation of aviation fuel. The Convention on International Civil Aviation, also known as the Chicago Convention, was signed in 1944 and has been revised eight times. It deals with the rules and regulations required to allow flights between countries. It also states that fuel, oil, spare parts, regular equipment and aircraft stores are exempted from any form of taxation, which means a carbon tax on aviation fuel to drive efficiency gains is currently not permitted. This is unfortunate, as not only can we build more efficient airplanes today, there are also alternative fuels that could be used. Biofuels could be developed as an additive or even as a replacement for traditional air fuel, kerosene. It is also possible to create artificial kerosene, by extracting carbon dioxide from the atmosphere and combing it with water. This takes a huge amount of energy but if electricity from renewable sources were used, it could be possible to have carbon negative aviation fuel. This would still only work if there were regulations or a carbon tax in place to make it cost effective to create artificial kerosene. In the short term, as there is no real fuel solution for aviation, the airlines are keen to be involved in carbon trading. This way, the airlines can 'off-set' their carbon emissions by ensuring an equivalent amount is saved elsewhere.

The other alternative is to persuade people to use public transport instead of their cars or flying[7]. For most people, it is clear that providing cheap and accessible electric buses, taxis, subway systems, and railways would all help reduce the number of car journeys. It can also help with freight and goods deliveries as the railway network can be used at night to transport goods around the country and between countries. Railways can also be used to replace internal and international flights. It has been calculated that all internal flights between American cities less than 600 miles apart could be replaced by high-speed electric 'bullet' trains travelling over 200 miles per hour, providing a quicker, safer and cleaner way to get around. High-speed trains running up and down the East and West coasts of the USA — with connections to the two major hubs of Chicago and Atlanta would remove 80% of the flights within the USA[8]. This sort of high-speed train network already exists in Japan, South Korea, as well as parts of China and the EU. There are also important high-speed rail networks being built around the world. For example, the Kuala Lumpur–Singapore high-speed rail (HSR) will link Kuala Lumpur in Malaysia with Woodlands in Singapore which should be fully opened by 2031. The trains will travel at 320 km/h (200 mph), cutting travel time between Kuala Lumpur and Singapore to about 90 minutes, removing the need for flights between the two countries and the huge network of buses in Malaysia. In addition, it would also reduce the need for both Malaysians and Singaporeans to drive between the two countries. It is clear that we need more of these high-speed rail connects around the world.

Fossil Fuel Subsidies

One of the major political problems with reducing carbon emissions concerns energy subsidies[9]. Firstly, there are huge fossil fuel subsidies, which continue to make oil, gas, and coal relatively cheap. Secondly, there is resistance to providing subsidies and tax incentives to the energy companies to build and supply renewable energy at competitive rates. A recent report from the IMF suggests that the fossil fuel industry receives over $7 trillion per year in subsidies[10] (two and half times the size of South Korea's annual GDP) — this includes direct payments, tax breaks, reduced retail prices, and the cost of climate change damage. Of this amount, governments provide at least $775 billion to $1 trillion as subsidies and at least $444 billion per year of direct funding to oil, gas, and coal companies to support exploration, extraction, and development. There is also a tremendous security cost associated with fossil fuels. A large part of foreign policy and military strategy for many countries involves protecting shipping lines for fossil fuels. The US military spends at least $81 billion a year protecting oil supplies. In comparison, there are no aircraft carriers defending wind turbine supply chains or strategic silicon reserves for solar panels.

The IMF report shows that fossil fuels account for 85% of all global subsidies and that they remain a large part of domestic policies. Had nations reduced subsidies in such a way as to create efficient fossil fuel pricing in 2015, the IMF believes that it "would have lowered global carbon emissions by 28% and fossil fuel air pollution deaths by 46%, as well as increased government revenue by 3.8% of GDP." It seems fossil fuel subsidies are bad for both the environment and the economy.

So why do fossil fuel subsidies persist? This may be down to the ownership of the major oil and gas companies. Out of the top 26 oil and gas companies, only 7 are private companies; the other 19 are fully or partly owned by countries. Hence, the state-owned companies are making huge amounts of money for the country and will continue to be given state aid in the form of subsidies and tax breaks to ensure that they are competitive with other oil and gas producing nations. This is only set to get worse with fracking and the shale gas revolution, with many countries such as the USA having found new reserves of natural gas underground.

Carbon Trading, Taxation, and Offsetting

There are three main policy approaches that can be used to help reduce carbon net emissions[9]. The first way to reduce carbon emissions is to impose a carbon tax on activities and goods that emit a large amount of carbon. Most economists agree that carbon taxes are the most efficient and effective way to curb emissions, with the least adverse effects on the economy. To avoid these taxes being regressive, the income from the carbon taxes should be used to support the least well off in society who will be the worst affected. Carbon taxes have been implemented in 25 countries, while 46 countries put some form of price on carbon, either through carbon taxes or emissions trading schemes.

The second approach is carbon trading, whereby carbon emissions are limited by the issuing of carbon permits. Carbon trading can drive innovation and push down costs. It is also a way of making renewable energy and Carbon Capture and Storage economically viable. Some emissions trading schemes allow companies to buy carbon offsets either nationally or internationally to count against their total emissions. Examples of mandatory compliance-based carbon trading schemes include: the European Union Emissions Trading System (EU ETS), China's national ETS, and various national and sub-national systems in countries like the US (RGGI), Canada, South Korea, and New Zealand. There are also industry and/or sector-based trading systems such as the new CORSIA (Carbon Offsetting and Reduction Scheme for International Aviation), which is a carbon market specifically for airlines to reduce emissions.

A carbon offset is defined as a reduction in emissions of carbon dioxide or other greenhouse gases made in order to compensate for emissions made elsewhere. This can be done either by increasing carbon storage through reforestation programmes or by retiring or removing emissions, for example closing down a coal-fired power station. There are two main carbon offsetting systems: the UN Clean Development Mechanism and the voluntary carbon markets (VCMs). The UN Clean Development Mechanism was a carbon finance scheme under the Kyoto Protocol (see Chapter 10) that allowed developed nations to fund emission-reduction projects in developing countries and receive credits to help meet their own climate targets. However, there were many challenges, including a limited number of countries participating and crash in credit prices, that have led to it being perceived as a failure and countries call for its end.

VCMs are, in comparison, conceptually simple and potentially nimble. They already channel \$2 billion dollars a year through their system, with the expectation that this could grow significantly over the next decade. The belief is that VCMs will also draw public funding into jointly funded mitigation (emissions reduction and removal) efforts. There are significant issues with the VCMs on the back of many media reports showing that offset projects have either not happened or have not significantly stored carbon. VCM problems include an oversupply of carbon credits because of the lack of retirement of old, now outdated, projects. There is also a lack of monitoring and verification of many projects, creating low confidence in the markets. In 2024, I was the lead author for the Climate Crisis Advisory Group report into VCMs which came up with seven recommendations to improve the markets and ensure it was effective in removing carbon from the atmosphere[11].

1. Transparency, including financial transparency:
   Every aspect of VCMs need be strengthened if transparent processes and calculations become the norm. In a transparent system, measurement and methodology become open for discussion, and as industry norms are strengthened, opportunities for 'shopping around' by buyers and sellers into VCMs also become available. Transparency about financial allocations along the VCM pipeline will help to expose unethical practices, and secure a fair deal for communities and project implementers.

2. Improve Monitoring, Reporting, and Verification
   MRV (Monitoring, Reporting, and Verification) is fundamental to carbon pricing and management mechanisms, supporting transparency and integrity by ensuring emission reductions are real and verifiable. These principles underpin all other improvements to VCMs because they enable transparency and meaningful scrutiny. MRV processes should progress towards standardisation wherever possible, and always be conducted in an open and transparent manner.

3. Do-no-harm credits and appropriate co-benefits: taking a holistic approach
   Climate change mitigation efforts, including carbon trading, can unintentionally lead to harm, or even to human rights breaches, especially

in the lives of indigenous people and local communities whose lives and livelihoods are intertwined with the resources under consideration. To avoid such unintended consequences, trustworthy carbon credit projects must be governed by 'do-no-harm' principles, ensuring they not only avoid causing harm, but also codify and record appropriate co-benefits.

4. Emissions removal is preferred to emissions reductions or avoided emissions
   In an ideal world, carbon removal from the atmosphere is to be preferred as opposed to emissions reductions (which should happen anyway, as part of a business commitment to net zero), or to 'avoided emissions' (where measuring and counting becomes even more difficult, since it depends on assumptions about what would happen to a carbon store, such as a forest, if no project was in place).

5. Carbon credits should eventually be allowed only for residual emissions
   Carbon credits should only serve as a tool for compensating a company's residual carbon footprints if all feasible measures to reduce emissions have been taken. The concept of net zero emphasises the necessity for comprehensive emissions abatement across all operational sectors, everywhere along the value chain, and anywhere in the world. Carbon credits, therefore, should not be seen as a primary strategy for achieving emissions targets, but rather as a supplementary measure, used only after a company has implemented all possible reduction technologies and strategies. This ensures that carbon credits contribute genuinely towards achieving global climate goals, rather than allowing companies to bypass the substantial reductions required under net zero commitments.

6. Increase demand for 'high quality' credits
   To stimulate demand for high-quality carbon credits, where MRV and financial transparency are given prominence, a robust framework of incentives for companies should be implemented, preferably through targeted government interventions and regulatory mechanisms. These

interventions could include tax benefits, subsidies, or direct regulatory requirements that promote the acquisition of premium credits over lower-quality alternatives. A comprehensive pricing mechanism is a crucial component of this approach and would ensure that companies opting for less expensive, lower-quality carbon credits, would have to purchase a greater quantity to achieve an equivalent environmental impact when compared objectively with the higher-quality (more expensive) credits. A rating system to identify high-quality credits and grade others would further enhance market transparency, build trust, simplify decision making and encourage companies to invest in superior options.

7. Government/International support and regulation
   There is a need for strong political support for regulation, transparency and governance of the VCMs at national, regional and international levels. Effective support could establish and enforce regulations that guarantee carbon credits are contributing effectively to emission reductions in accordance with the host countries' Nationally Determined Contributions (NDCs). Strengthening political infrastructure in relation to VCMs will enhance the market's credibility and ensure that carbon offsetting schemes contribute meaningfully to global emission reduction targets over the long term, making them an integral part of a global strategy for climate action.

Reforestation and Rewilding

One of the most important ways to remove carbon dioxide from the atmosphere is through reforestation and rewilding[3]. Since the beginning of agriculture, it has been estimated that humans have cut down three trillion trees — about half the trees on Earth. We know that the Earth can sustain a much larger forested area. Rewilding habitats and reforesting may be easier in the future as the world is already becoming a wilder place. This may seem counter intuitive, given that the global population will grow from over 8.2 billion today to 10 billion by 2050, but by then, nearly 70% of us will live in cities and we may abandon many remote rural areas, making them ripe for restoration. Already, in South Korea, forest area has increased by over 3 million hectares from the mid-1950s to 2007, which is an approximately 30% increase of forest cover. In Europe, 2.2 million hectares of forest has regrown per year between 2000–2015. In Spain, forest cover has increased from 8% of the country's territory in 1900 to 25% today, while in the UK, forest cover hit a low of 5% after the First World War and is now back up to 13%.

Massive reforestation is not a pipe dream; it can have real benefits for people. In the late 1990s, environmental deterioration in western China became critical, with vast areas resembling the Dust Bowl of the American Midwest in the 1930s. Six bold programmes were introduced, targeting over 100m hectares of land for reforestation. Grain for Green is the largest and best known of these. These radical tree-planting programmes had an amazing effect as the trees stabilised the soils, greatly reducing soil erosion and the impacts of flooding. Through transpiration, the trees added moisture to the atmosphere, reducing evaporation and water loss. Once the forests reached a critical size and area, they also started to stabilise the rainfall. All these impacts combined to boost local agricultural production. The ongoing programme has also helped to alleviate poverty as direct payments were made to farmers to set aside their land for reforestation. This was an excellent example of the win-win solution required for climate change as it increased carbon storage, improved the local environment, and helped alleviate extreme poverty.

In 2019, researchers claimed in the journal *Science* that covering 900m hectares of land — roughly the size of the continental US — with 1 trillion trees could store up to 205 billion tonnes of carbon[12], about two thirds of the carbon that humans have already put into the atmosphere. The 'one trillion trees' mantra has caught the popular imagination and even President Trump declared at the Davos meeting that this was a good idea. The only problem is that the research that produced this high carbon number was fundamentally flawed[13] and in fact, the IPCC and other studies suggest that new forests could store on average an extra 57 billion tonnes of carbon by the end of the century. This is still a large number, but considering we put 11 billion tonnes of carbon into the atmosphere per year, this represents only 5 years of human pollution. Therefore, it is evident that reforestation alone is not an alternative to rapid and deep cuts in our fossil fuel emissions. However, we will require negative carbon emissions to keep the global warming to 1.5°C later in this century and one of the key ways of creating negative emissions is through reforestation.

Already 61 countries have joined the Bonn Challenge and pledged to restore 350m hectares of degraded land to forest worldwide by 2030. That is an area fifteen times the size of the UK. But there is another issue. Massive reforestation only works if the world's current forest cover is maintained and increased. As noted earlier, deforestation of the Amazon rainforest and many other forests around the world is still ongoing. Current estimates suggest that areas of rainforest the size of a football pitch are being cleared every single minute.

Reforestation and afforestation are fundamentally limited by the land area available, as trees can only hold a finite amount of carbon. We must also remember that reforestation is not always the best option, and this is why the term 'rewilding' is used in conjunction with reforestation. For example, draining wetlands or peatlands to plant forests is counter-productive, as the carbon storage will be lower and losses of biodiversity will be considerable. Hence, the most appropriate restoration project must be applied in each region of the world. This could be re-wetting wetlands, preserving peatlands, re-growing mangrove forests, or maintaining open grasslands. If an area is suitable for reforestation, decisions have to be made about the most appropriate species for that area in terms of current and future climate and how to increase local biodiversity and other ecological services. One of the criticisms of the Bonn Challenge is that about half the pledges involve large-scale commercial forest plantations. Plantations only lock up carbon while the trees are growing, and much of this is returned when they are harvested. In any case, monoculture is bad for biodiversity as well.

Summary

If we are to reduce the impacts of climate change, we need to reduce the amount of greenhouse gas pollution that we are emitting to net zero while enabling the very poorest countries to develop in a sustainable way. The world's population is currently just over 8.2 billion, and it is likely to rise and plateau at 10 billion by 2050. That adds up to eight billion people who would like the same lifestyle as a person living in the developed world, which could potentially mean a huge increase in greenhouse gas emissions in this century to fuel this consumer dream if they follow the same development pathway. At the very least, we need to lift over 2 billion people out of extreme poverty which requires massive expansion of energy to these populations. The good news is we already have a lot of the technology needed to mitigate greenhouse gases emissions over the next 70 years. Renewable energy, energy efficiency, CCS, carbon trading, offsetting, reforestation, and rewilding all have a role to play. Many of these technologies have added bonuses such as reducing air pollution, decreasing energy prices and making energy more secure. To pay for this decarbonisation of our global economy would take between 1–3% of the world's GDP, to offset a much larger bill in the future. But it would mean shifting profits from the fossil fuel industry to new 'disruptive technologies' and of course, this is being met with huge resistance. Mitigation is at a fundamental level the painful messy transition from 20th century technology which involved burning stuff, to 21st century technology which generates and utilises electricity in a much clean, efficient and safe way.

Chapter 12

Solutions: Adaptation

Introduction

There have already been climate change impacts, and these will increase as the global temperature continues to rise. The second report of the IPCC Sixth Assessment examines the impacts of climate change as well as the potential sensitivity, adaptability, and vulnerability of each national environment and socioeconomic system[1]. The IPCC gives six clear reasons why we must adapt to climate change: (1) climate change impacts cannot be avoided even if emissions are cut rapidly to zero; (2) anticipatory and precautionary adaptation is more effective and less costly than enforced last-minute emergency fixes; (3) climate change may be more rapid and pronounced than current estimates suggest, which leads to unexpected and extreme events occurring; (4) immediate benefits can be gained from better adaptation to climate variability and extreme atmospheric events (for example, with storm risks, strict building laws and better evacuation practices would save lives); (5) immediate benefits can also be gained by removing maladaptive policies and practices (for example, building on floodplains and vulnerable coastlines); and (6) climate change brings opportunities as well as threats.

The major threat from climate change is its unpredictability. As noted earlier, humans live within a huge range of climate, from baking deserts to the frozen Arctic, but we only do so because we can predict what the extremes of the weather we must cope with. As climate change impacts increase, the weather will become more extreme and unpredictable. There are four types of adaptation strategies: infrastructural, institutional, behavioural and nature-based solutions.

Infrastructural Adaptation

With increased extreme weather events and rising sea levels, physical adaptations are required which means thinking about infrastructure changes[2]. This may mean building or upgrading existing infrastructure to protect against storms, floods, droughts, heatwaves, extreme heat and sea level rise. This also includes protecting transport networks such as roads, railways, airports and ports from extreme conditions because transport enables essential goods and services such as food, fuel, people, and medical supplies to move to where they are needed. It is also imperative that the countries or region's energy network is protected. This includes power stations, nuclear power stations, wind turbines, solar farms/panels, hydroelectric dams, biofuels and geothermal. There are some crucial questions that Governments around the world need to consider. For example, are power stations protected from extreme events and do they have enough water supply to generate electricity? How is electricity being distributed within the country and is this transmission secure? Is there sufficient protection for the fresh water supply that is essential for people, agriculture and industry? Each of these questions need to be addressed and changes made to improve the infrastructure. For example, in many countries, electricity is distributed by overhead cables which are vulnerable to high winds. The Governments of these countries may need to rewire their electricity distribution networks to combat against extreme weather events and provide redundance in case of failure. With increased droughts and heavy rainfall events, Governments also need to upgrade the infrastructure used to collect and store water to ensure an adequate freshwater supply.

Cities are uniquely vulnerable to climate change as a large number of people live closely together[3]. Each city has its own set of risks and local Governments need to devise their own adaptation plans. Sea level rise will affect coastal communities and ports so Governments need to review which parts of the coastline are vital and need to be protected and which

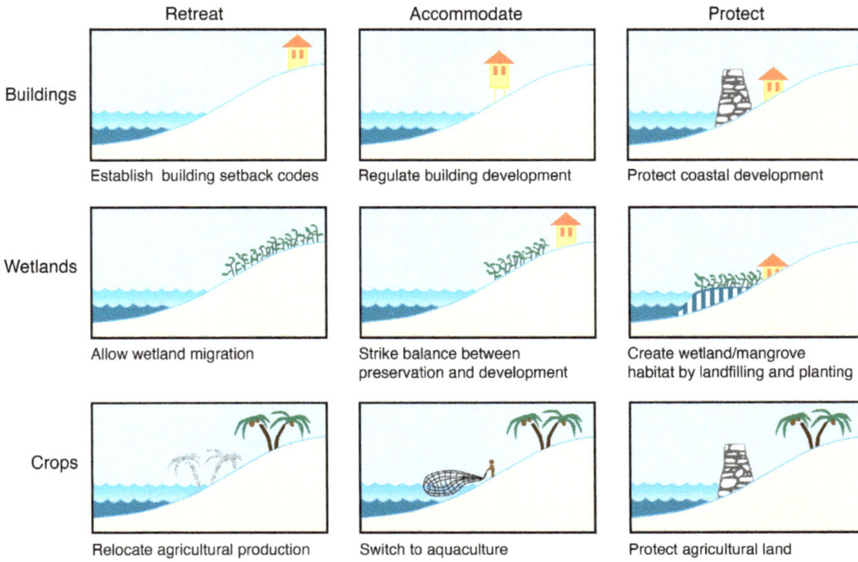

Figure 54. Alternative Strategies for Dealing with Future Sea Level Rise

can be allowed to retreat. Ultimately, a holistic approach to coastal management is required as hard physical infrastructure may not always be useful while nature-based solution (discussed below) and managed retreat may be more appropriate (see Figure 54).

Infrastructure changes are costly and take time. In many countries, these large infrastructure projects can take up to 30 years to plan, develop, and build. If we take the issue of combating sea level rise (Figure 55), the potential timeline can look like this: ten years to research and plan the appropriate measures, ten years for the full consultative and legal processes, and a further ten years to implement these changes. It can then take another decade for the natural restoration to take place to complete the sea level adaptation project. A good example of this is the Thames Barrier, which currently protects London from floods, It was developed in response to the severe flooding in 1953, but did not open officially until 1984, 31 years later. This is why we need institutional adaptation as planning and building of new infrastructure needs to be on the same time scale as rapid climate change — which is happening in years, not decades.

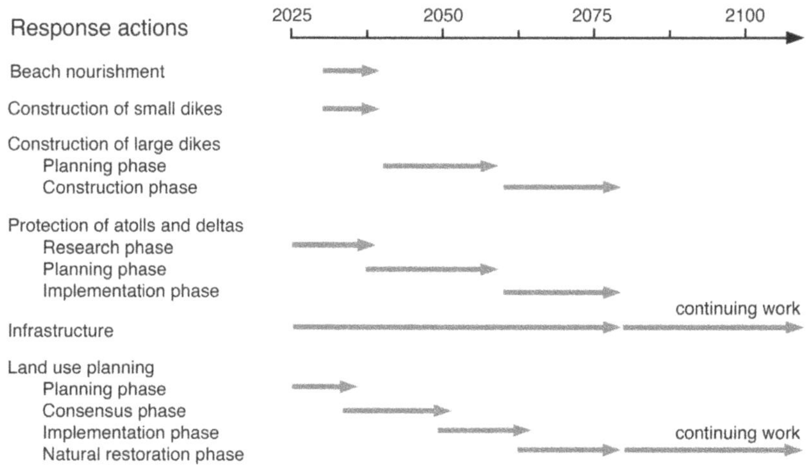

Figure 55. Lead Times for Response Strategies to Combat Climate Change

Institutional Adaptation

All national and local Governments need to have a climate change adaptation plan[4]. Many do but there is a massive gap between planning and implementation — the barriers to action are discussed below. Institutional responses are wide ranging and can include: new infrastructure projects, new planning and zoning regulations, new building regulations, new insurance schemes, and coordination mechanisms between local and national Governments as well as the private sector. One interesting case study is insurance, which helps to spreads the financial impact of flooding and other extreme weather events. Many Governments support or underwrite insurance schemes, which is essential as most private insurance companies wish to avoid higher risk areas and clients. Government schemes spread the cost of climate change impacts across the whole population and time, while supporting private insurers operating in a competitive market. However, these schemes must be combined with strong planning and building regulations or there could be perverse incentives to develop properties in high-risk areas. National and local Governments can also communicate with the public to highlight the safety and health risks of the changing climate. Hazard early warning systems are essential to protect people and save lives. In Chapter 7, this was illustrated by the changes in Bangladesh that have led to a massive drop in fatalities from cyclones.

Behavioural Adaptation

Individuals, households, civil society and the private sector all have a central role to play in adaptation[2]. Some changes can occur slowly, such as warmer summers or wetter winters which allow time for individuals and organisations to adapt to new conditions. However, some changes are extremely fast, for example, an extreme weather event such as a massive flood or major heat wave. These require organisations and individuals to shift behaviour abruptly. A good example of the latter is the 2003 European heat wave. Over 75,000 excess deaths occurred due to this heat wave — many of these deaths were in France — so the French Government completely reassessed its health response to the crisis. They changed everything, including communication with the public, having a registry of vulnerable individuals, local health responses, and hospital admissions and treatment. It is estimated that in subsequent heat waves, the death toll was cut by over 75% because of these quick and decisive social adaptations[5].

Agriculture offers many opportunities for behavioural adaptation. These include changing planting times, or changing to crops and livestock that are better adapted to new climate conditions as well as the presence of new pests and diseases. Other examples are breeding more resilient crops and selecting suitable genetically modified crops. These adaptations should be combined with changes in individual diets — moving away from meat towards more plant-based diets. As well as having a much lower carbon footprint, plant-based options have much lower energy and water requirements. All these adaptations will improve food security and nutrition. In many countries, food and water security will become the major issue, and therefore, policies and behaviour changes are essential to safeguard people's access to food and clean water even when they are unable to pay. In many ways, the most important adaptation to climate change is good governance, so that policies can be formulated and enacted to protect the most vulnerable people in society[6].

One of the most controversial behavioural changes is migration. Migration is a complex topic and whole books have been written on it. There are many factors influencing migration decisions: economic, demographic, political, and conflict. It is difficult to say how much climate change has or will influence migration, but as discussed in the security section, climate change is seen as a threat multiplier. Climate change is impacting agriculture and making incomes and employment less certain. And this is contributing to the rural to urban migration.

Nature-Based Adaptation

Nature-based solutions are defined as the use of nature and natural processes to address issues such as water and food security, disaster risk reduction, as well as both climate change mitigation and adaption[2]. For example, nature-based systems for climate change adaptation can include natural flood management such as restoring natural coastal defenses, and providing local cooling.

Everyone is familiar with the need to preserve tropical rainforests (discussed in Chapter 11) and to reforest appropriate areas of the world. However, there are also other pristine habitats, on land and in the ocean, which are in dire need of protection. Mangrove swamps occupy less than 1% of Earth's surface, but store the equivalent of 22 billion tonnes of carbon dioxide. That is around two-thirds of total emissions from burning fossil fuels each year. These coastal habitats act as a home, nursery, and feeding ground for numerous species. More than forty bird, ten reptile and six mammal species are only found in mangroves. Peatlands — those soggy ecosystems which include bogs, marshes and fens — store twice as much carbon as all the world's forests. The top 15 cm stores more carbon below ground than tropical rainforests do above ground. In the UK, peatlands store the equivalent of ten billion tonnes of carbon dioxide and host precious plant and animals such as red grouse, mountain hares and marsh ear wort. Both mangrove swamps and peatlands are extremely good at preventing floods. Mangroves swamps moderate the impact of waves and wind on coastal settlements or cities while peatlands absorb a lot of water and release it slowly to prevent flash floods.

It is possible to manage farmland and fisheries sustainably. Not all the world's land and ocean can be left to nature, as we need the land and ocean to produce food and other resources. However, we can be managed them better. People currently use about 25% of the planet's land surface for growing food, extracting resources and living. The global food system contributes one-third of all greenhouse gas emissions. It is possible to improve and modernise methods of farming, for example, agroecology, which involves incorporating trees and habitats within farm fields. It is possible to use sustainable fishing practices which can help protect and regenerate topsoil and seabed habitats, boosting biodiversity and improving the resilience of these ecosystems to climate change.

Another set of nature-based solutions involved urban green roofs, blue roofs and green walls — which can also be included under infrastructure. They can reduce the effects of urban heat islands, capture stormwater, abate pollution, and act as carbon sinks. At the same time, they can enhance local biodiversity.

Stabilising the Population

A frequent suggestion is that we need to stabilise the global population, as a way of reducing emissions and the burden on adapting to climate change[7]. A process of what is known as the demographic transition has led global population to increase by more than double since the 1970s. In the earlier stages of a country's development, societies tend to have high child mortality rates that are offset by high fertility rates, leaving the population relatively stable. Couples have as many children as they can to ensure that some — on average two — survive to adulthood. As societies develop a more stable food supply, better sanitation and widely available treatment for common diseases, mortality rates drop rapidly while fertility rates stay high, resulting in a rapid population growth. Depending on the time lag between decreasing mortality and stable fertility, the post-transition population can be between four and ten times higher than pre-transition.

Many people assume that universal access to family planning and the availability of contraception are key to reducing fertility rates quickly, but the first demographic transition can be traced to the European

Enlightenment, just before the 19th century — when these services were not available. Instead, it is women's education up to and beyond secondary school level that is the critical control of fertility. The rapid spread of vaccinations and the expanded food supply created by the green agricultural revolution in the 1960s meant that at its peak, global population was growing at over 2% per year. In 1950, each woman gave birth on average to five live babies. Now that the demographic transition has already happened in many countries around the world, the figure is below 2.5. While the average birth rate is falling every year, the world's population is still growing by 200,000 a day. The United Nations predicts that the population will rise to approximately 9.7 billion by 2050, and it will stabilise around 10 billion by 2100[8]. That is another 1.5 billion people.

The first argument about population is that we have too many people to feed and that is the reason holding back climate change adaptation. This is a false narrative — according to the World Food Programme, there are 730 million people on the brink of starvation today, but the World produces enough food to feed 10 billion people — easily enough to cover even the predicted increase in population this century. As discussed in Chapter 7, the reason there are so many people starving is because they cannot access this global food surplus, usually through the lack of money. When the very poor lose their livelihoods through civil unrest, war or crop failure, they have no resources to fall back on, and no money to buy the food they need to survive.

The second false argument is that stabilising population is a key solution to climate change as it will reduce emissions. This is misleading and unhelpful because it makes a simplistic assumption that everyone's contribution is equal. A third of the carbon pumped into the atmosphere to date has come from the US, and another third from the EU. Africa has contributed just 3%. Over-consumption by the wealthy is causing climate change, not rising population. Put another way, the average American emits nine times more CO_2 than the average Indian — so, in line with this argument, population reduction in the US would be much more effective in curbing GHG emissions than stabilising growing populations elsewhere. Some argue that new populations will eventually emit more as societies continue to develop. However, the climate crisis is now and the world needs to go carbon neutral by 2050. New populations and societies are therefore, not relevant.

Nonetheless, having a stable population is also essential at the country level. It allows governments to better ensure food, water and resource security for all their citizens. It also makes it easier to provide improved healthcare services and economic opportunities for a greater proportion of citizens. An excellent example is the huge challenges that Nigeria faces with a population that has grown by 100 million people in less than 20 years. Women's education and universal access to family planning around the world will help stabilise population, and bring major benefits such as resilience and adaptive capacity in areas that are most vulnerable.

Adaptation and National Security

Climate change has become a major concern for national security as it impacts the health, resilience and economy of a country. It can also affect global stability through severe weather events, mass migration, as well as increasing food and water insecurity. It can be seen as a 'threat multiplier'; while climate change itself does not necessarily cause instability or conflict, it can increase social and political tensions and drive regions and countries into instability.

A recent report by the UK Alan Turing Institute suggested that climate security risks can be described in three broad categories[9] — physical risks, human mobility and social cohesion risks, as well as transition risks. Physical risks refer to resources, critical infrastructure and people. These include threats to water, sanitation, energy, transportation, food production and human health. Human mobility and social cohesion risks include increase poverty, migration, displacement as well as potential conflict and violence within and between countries. Transition risks are security risks associated with establishing a low-carbon global economy as changes to global markets, technology, and trade policies can all greatly affect a country's economy and thus its security.

Every country needs to assess the impact of climate change firstly on their own internal security and ensure there is an adaptation plan to protect energy, water, transportation and food production. Next, countries need to

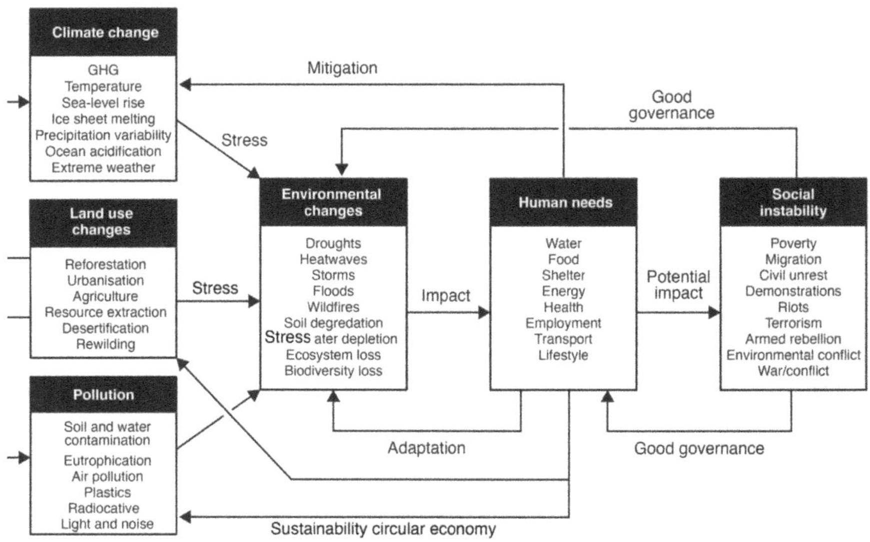

Figure 56. Climate Change Impacts on National Security

acknowledge that climate change is a threat multiplier and will make the world less secure unless steps are taken to reduce global emissions as fast as possible and adaptation is supported properly, especially in the most vulnerable countries and regions.

Barriers to Adaptation

Adaptation seems such a sensible thing to do, so why are countries not adapting as rapidly as they should?[2] There are a number of major barriers that have been identified. The first problem is that adaptation requires money to be invested now; many countries just do not have the money, and in countries that do, people are unwilling to pay more taxes to protect themselves in the future, as most people live for the present. In 2024, the United Nations Environment Programme estimated the amount of money the Global South need to spend per year to adequately adapt to climate change — it was a staggering $187 billion to 359 billion[10]. The second problem is the delay in action despite many countries having adaptation plans because in many cases, we are unsure what climate changes are going to occur and when. Moreover, adaptation is seen as long term, and there are always more seemingly pressing short-term issues. This is compounded as many countries also have short election cycles of 4 to 5 years, which means politicians are always thinking short-term and rarely about the longer-term. This is, of course, despite the fact that all of the adaptations discussed will in the long term save money for the local area, the country, and the world. The third issue is that the global economy is dominated by markets which are set up to create short term gains, and have few, if any, mechanism to promote longer term strategic. Additionally, Governments and international institutions seem reluctant to impose strong regulation despite the financial crash of 2008 due to failure of the markets. The fourth issue is that developed countries are reluctant and have failed again and again to provide the financing required by the least developed countries to adapt. Even when money is provided, it is short-term and usually has to be spent on companies from the donor country to boost their own economy instead of developing long-term adaptation for the recipient country. The last issue is that there are some climate change impacts that are so large and catastrophic that countries simply do not have the money to prevent, avoid or adapt to[12].

Summary

There are limits to adaptation. In some regions, climate change impacts may become so large that they go beyond our ability or finances to protect the population living there. Continued sea level rise will mean many small island nations may become uninhabitable eventually and the population will have to relocate. In 2019, Indonesia's president Joko Widodo announced that the national capital will move from Jakarta, on the island of Java, to the province of East Kalimantan, on Borneo. This was in part to relieve pressure on the capital and address inequality in Indonesia, but it is also because Jakarta is sinking. Areas of north Jakarta, including the sea wall designed to protect the population, are falling at an estimated 25 cm a year, due to sea level rise and the extraction of fresh water from shallow aquifers, leading to subsidence.

As climate change warms up the planet, the average annual temperature of each region will increase significantly. If we take the worst-case scenario in climate change (see Chapter 6), this mean 3.5 billion people theoretically would have to move if they wanted to remain under the same climate range as today. Even if strong climate policies were to keep global temperature increase to 2°C, there could still be at least 1.5 billion people who would theoretically have to move to maintain their current climate. Now this mass migration is highly unlikely, but it shows how much adaptation will be required to safeguard people even with just a 2°C warming.

Chapter 13

Solutions: Geoengineering

Introduction

Geoengineering is the general term used for technologies that could be used to either remove GHGs from the atmosphere or to change the climate of the Earth (see Figure 57)[1]. Ideas considered under geoengineering range from the very sensible to the completely mad. It is crucial to note that geo-engineering is not an alternative to massive reductions in global GHG

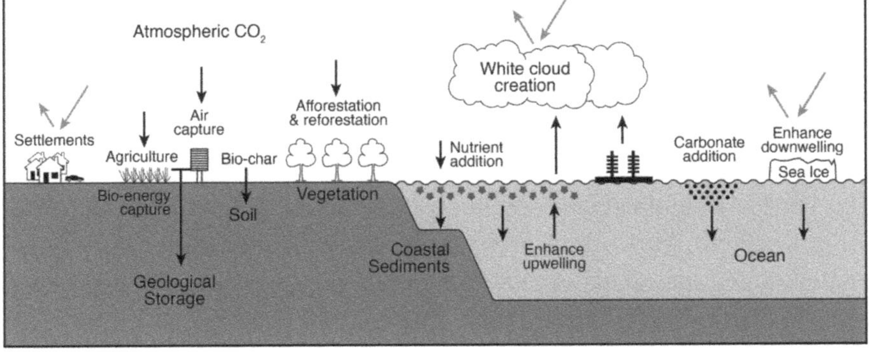

Figure 57. The Range of Geoengineering Approaches

emissions. Instead, most people see geoengineering solutions as a fall back if we are unable or unwilling to cut GHG emissions quick enough.

Carbon Capture and Storage (CCS)

Removal of carbon dioxide during industrial processes can be tricky and costly, because the gas needs to be removed and stored somewhere[2]. The IPCC special report on 'Carbon dioxide capture and storage' published in 2005 concluded that the technology for CCS existed, but there is little commercial experience in configuring all of the components needed to create fully integrated CCS systems at the kinds of scales needed in the future. Power production costs that include CCS would rise by at least 15% and could be as high as 100%. Many argue that the plus point for CCS is that some of the recovered carbon dioxide may be utilised in enhanced oil recovery, the food industry, chemical manufacturing (producing soda ash, urea, and methanol), and the metal-processing industries. Carbon dioxide can also be applied to the production of construction material, solvents, cleaning compounds, and packaging, as well as in waste-water treatment. However, in reality, most of the carbon dioxide captured from industrial processes would have to be stored. It has been estimated that theoretically two-thirds of the carbon dioxide formed from the combustion of the world's total oil and gas reserves could be stored in the corresponding reservoirs. Other estimates indicate storage of 90–400 Gt in natural gas fields alone and another 90 Gt in aquifers.

Oceans could potentially also be used to dispose of the carbon dioxide. Suggestions have included storage by hydrate dumping: mixing carbon dioxide and water at high pressure and low temperatures creates a solid, or hydrate, which is heavier than the surrounding water and thus drops to the bottom of the ocean. However, with ocean storage, there is the added complication that the oceans circulate, so some of the carbon dioxide that is dumped will eventually be returned to the surface and then the atmosphere. Moreover, scientists are very concerned about the effects on the ocean ecosystems. Another more recent suggestion is to inject carbon dioxide half a mile deep into shattered volcanic rocks in between giant lava flows. The carbon dioxide will react with the water percolating through the rocks. The acidified water will then dissolve metals in the rocks, mainly calcium and aluminium. Once it forms calcium bicarbonate with the calcium, it can no longer bubble out and escape. If it does escape into the ocean, then bicarbonate is relatively harmless.

The major problem with all these methods of storage is safety. Carbon dioxide is a very dangerous gas because it is heavier than air and can cause suffocation. This was powerfully illustrated in 1986, when a large release of carbon dioxide from Lake Nyos, in the west of Cameroon, killed more than 1,700 people and livestock up to 25 km away[3]. Although similar disasters had previously occurred, never had so many people and animals been asphyxiated on such a scale in a single brief event. Scientists now believe that dissolved carbon dioxide from the nearby volcano seeped into the lake from springs below and was trapped in deep water by the weight of water above. In 1986, there was an avalanche that churned up the lake waters, resulting in an overturn of the whole lake, and an explosive release of all the trapped carbon dioxide. Nevertheless, huge amounts of mined ancient carbon dioxide are pumped around the USA to enhance oil recovery, and there have been no reports of any major incidents so far. Engineers working on these pipelines feel they are much safer than gas and oil pipelines, many of which run across most major US cities.

Carbon Dioxide Removal

There are three main approaches to the removal and storage of atmospheric carbon dioxide: biological, physical, and chemical[4].

Biological carbon removal: One of the most important ways to remove carbon dioxide from the atmosphere is through reforestation and rewilding. Since the beginning of agriculture, it has been estimated that humans have cut down three trillion trees — about half the trees on Earth. The Earth can, therefore, sustain a much larger forested area. Rewilding habitats and reforesting may be easier in the future as the world is already becoming a wilder place. This may seem a counter intuitive, given that the global population will grow from over 8.2 billion today to 10 billion by 2050, but by then, nearly 70% of us will live in cities and we will have abandoned many remote rural areas, making them ripe for restoration. Already, in Europe, 2.2 million hectares of forest has regrown per year between 2000–2015. In Spain, forest cover has increased from 8% of the country's territory in 1900 to 25% today, while in the UK, forest cover hit a low of 5% after the First World War and is now back up to 13%.

Massive reforestation is not a pipe dream; it can have real benefits for people. In the late 1990s, environmental deterioration in western China became critical, with vast areas resembling the Dust Bowl of the American Midwest in the 1930s. Six bold programmes were introduced, targeting over 100m hectares of land for reforestation. Grain for Green is the largest and best known of these. These radical tree-planting programmes had an amazing effect as the trees stabilised the soils, greatly reducing soil erosion and the impacts of flooding. Through transpiration, the trees added moisture to the atmosphere, reducing evaporation and water loss. Once the forests reached a critical size and area, they also started to stabilise the rainfall. All these impacts combined to boost local agricultural production. The ongoing programme has also helped to alleviate poverty as direct payments were made to farmers to set aside their land for reforestation. This was an excellent example of the win-win solution required for climate change as it increased carbon storage, improved the local environment, and helped alleviate extreme poverty.

In recent years, there has been a campaign to plant an extra one trillion trees — in terms of land needed, this would be equivalent to the whole of continental USA. The 'one trillion trees' mantra has even been supported by President Trump. These new forests could store, on average, an extra 57 billion tonnes of carbon by the end of the century. This is a large number, but considering we put 11 billion tonnes of carbon into the atmosphere per year, this represents only five years of human pollution. Therefore, reforestation is not an alternative to rapid and deep cuts in our fossil fuel emissions. Nonetheless, later in this century, we will require negative carbon emissions to keep the global warming to 1.5°C — one of the keyways of creating negative emissions is through reforestation.

The second biological approach is changing the uptake of carbon dioxide by the oceans. The most famous ocean 'technofix' was suggested by the late Professor John Martin[5]. He suggested that many of the world's oceans are under-producing because of the lack of vital micro-nutrients, the most important of which is iron, as it allows plants to grow in the surface waters. Marine plants need minute quantities of iron, and without it, they cannot grow. In most oceans, enough iron-rich dust gets blown in from the land, but it seems that large areas of the Pacific and Southern

Ocean do not receive much dust and thus are barren of iron. It has been suggested that we could fertilise the ocean with iron to stimulate marine productivity. The extra photosynthesis would convert more surface-water carbon dioxide into organic matter. Furthermore, when the organisms die, the organic matter drops to the bottom of the ocean, taking with it and storing the extra carbon. The reduced surface-water carbon dioxide is then replenished by carbon dioxide from the atmosphere. So, in short, fertilising the world's oceans could help to remove atmospheric carbon dioxide and store it in deep-sea sediments. In reality, the scientific evidence has been disappointing. Experiments at sea have produced highly variable results, with some showing no effects at all while others revealing that the amount of iron required would be huge. To compound this as soon as we stop adding the extra iron, most of this stored carbon dioxide is released back into the air. This is because surface ocean biology is very efficient and very little organic matter is allowed to escape out of the photic zone (the part of the ocean close enough to the surface to receive the necessary sunlight for photosynthesis to take place).

Physical. It is possible to remove carbon dioxide directly from the air. However, considering carbon dioxide makes up just 0.04% of the atmosphere, this is much harder and more expensive than it sounds. Nonetheless, technologies to remove carbon dioxide at source and from the atmosphere are being rapidly developed. For example, Climeworks technology uses giant fans to collect carbon dioxide directly from the air, producing pure carbon dioxide to use in industrial processes or even for making artificial fuels[6]. The issue is now one of cost and financing, not the engineering.

Chemical. Carbon dioxide is naturally removed from the atmosphere over hundreds and thousands of years through the process of weathering, at a rate of 0.1 gigatonnes of carbon per year, but this is 100 times smaller than the amount we are emitting. This is because rainwater dissolves some carbon dioxide as it passes through the atmosphere, making a very weak 'carbonic' acid. This is why many buildings in our cities made out of limestone seem to be wearing away overtime. The type of rock that is being weathered is important and determines whether the weathering removes carbon from the atmosphere. For example, weathering of carbonate rocks or limestone building simply dissolves the rock which creates carbon dioxide that goes straight back into the atmosphere. However, weathering of silicate rocks creates minerals such as bicarbonates (HCO_3^-), when these are washed into the oceans, marine plankton can use this to make calcium carbonate shells. These calcite shells are ultimately deposited at the bottom of the ocean in deep-sea sediments and hence the carbon is lost from the atmosphere. Now, on a geological timescale, these sediments will only store carbon as long as the oceanic crust is not recycled (see Chapter 4), but that will be tens of millions of years in the future.

There are a number of geoengineering ideas aimed at enhancing these natural weathering reactions. One suggestion is to add silicate minerals to soils that are used for agriculture. This would remove atmospheric carbon dioxide and fix it as carbonate minerals and bicarbonate in solution. The scale at which this would have to be done is very large and there are unknown effects on soils and their fertility. Another suggestion is to enhance the rate of reaction of carbon dioxide with basalts and olivine rocks in the Earth's crust. Concentrated carbon dioxide would be injected into the ground and would create carbonates deep underground. An example of this is the Icelandic ON Power geothermal park where CarbFix takes pure carbon dioxide provided by Climeworks technology and inject it into the underground basaltic rock formations. Geothermal renewable energy provides the energy for the direct air capture and injection of the carbon dioxide, and early estimates suggest that the system could permanently remove 4000 tons of carbon dioxide from the air per year. To put that into context, Iceland would need 1000 such plants to remove its current annual total carbon emissions. The positive side is that we now have a tried, tested, and safe CCS system.

Solar Radiation Management

Solar radiation management: Another approach that scientist have suggest is to cool the planet down using other methods instead to reducing atmospheric GHG level[1]. One way would be to reduce the amount of sunlight hitting or being absorbed by the Earth as that will potentially reduce the total energy budget and may result in a cooler Earth. As will be evident from the above, some geoengineering solutions are still just ideas and need a lot more work to see if they are even feasible. This is particularly true of the solar radiation management ideas — many of which sound like something out of a bad Hollywood B-movie. These suggestions include changing the albedo (reflectivity) of the Earth to increase the amount of solar energy reflected back into space to balance the heating from global warming. The various ideas to increase the albedo include erecting massive mirrors in space, injecting aerosols into the atmosphere, making crops more reflective, painting all roofs white, increasing white cloud cover, and covering large areas of the world's deserts with reflective polyethylene-aluminium sheets.

The fundamental problem with all of these approaches is that we have no way of predicting their overall effect on climate. Let us examine the mirrors in space idea put forward by Roger Angel, Director of the Centre for Astronomical Adaptive Optics at the University of Arizona. First and foremost, it would be expensive, requiring 16 trillion gossamer-light spacecrafts costing at least $1 trillion dollars and taking 30 years to launch[7]. Second, like all the geoengineering ideas to change the Earth's albedo, it may not even work the way we want it to. These approaches are aimed at getting the Earth's average temperature down, but they may end up changing the distribution of temperature with latitude, which is what drives climate (see Chapter 1). Some climate models have shown that these approaches may give a different global climate, with the tropics being 1.5°C colder, the high latitudes 1.5°C warmer, and precipitation varying unpredictably around the world.

Governing Geoengineering

One of the major issues with geoengineering is how to govern different groups, companies, and countries playing with the global climate system[1]. There are a great many ethical issues that arise when considering how changing regional and global climate may affect countries differently[8]. There may be overall positive results, but minor changes in rainfall patterns, which could mean that some countries will receive too little or too much rain, resulting in disaster. There are three main views on geoengineering: (1) it is a route to buying back some time to allow the UNFCCC negotiations to catch up with need to achieve net zero carbon by 2050[1]; (2) it represents a dangerous manipulation of the Earth system and may be intrinsically unethical[9]; and (3) it is strictly an insurance policy to support mitigation and adaptation efforts if they fail to be sufficient on their own[2]. Even if research is allowed to go ahead and geoengineering solutions are required, like many emerging areas of modern technology, new flexible governance and regulatory frameworks will be required. Currently, there are many international treaties with a bearing on geoengineering, and it seems that no single instrument applies. Hence, geoengineering, like climate change, challenges our nation-state view of the world, and new ways of governing will be required in the future.

Summary

We are currently undertaking the biggest experiment on the Earth system by injecting huge amounts of GHGs into the atmosphere which is warming up the planet and changing our climate. The logical conclusion would be to stop emitting so much GHGs, but as we have seen in early chapters, there is political and social inertia that is making this transition slower than needed. Hence, geoengineering may offer a way to buy us time to get to net zero global emissions. First, there are processes that can capture and remove carbon dioxide from our industrial processes and the atmosphere. The critical questions here are the costs, who pays them and how the carbon dioxide is safety stored. Some of these seem very sensible as they have other positive impacts on our environment — such as massive reforestation which can stabilise soils and rainfall while reducing the threat of flash flooding. Second, there is solar radiation modification, whereby some scientists suggest it may be possible to reflect sunlight away from the planet to cool it down. This seems like wanting to try another massive experiment to cover up the impact of the first inadvertent experiment. There are huge scientific, political and ethical issues with solar radiation modification. The simple conclusion is it is safer, cheaper and simpler just to reduce GHG emissions around the world.

Chapter 14

Changing Our Future

Introduction

The challenge of climate change must be seen within the current dominant political and economic landscape. Only by understanding the fundamental societal and economic causes of carbon emissions can we hope to be able to build systems that can rapidly reduce them. At the same time as we deal with climate change, we need to ensure that we also tackle other global challenges, such as poverty and inequality, environmental degradation, and security. Future policies and international agreements need to provide win-win solutions that deal with the biggest challenges facing humanity in the 21st century.

Planetary Stewardship

Scientists struggle with the fact that despite the huge weight of evidence for climate change, a small, but significantly vocal minority of influencers deny that climate change is happening. Scientists have responded by collecting even more evidence. This is called the 'deficit model', in which scientists perceive that decisions are not being taken due to a lack of information. However, social scientists have found that acceptance of climate change has little to do with science and everything to do with politics. Climate change challenges the Anglo-American neoliberal view that is held by many mainstream economists and politicians[1]. It shows a fundamental failure of the market and requires governments to act collectively to regulate industry and business[2]. It is one of the greatest ironies of climate change that the very politicians who are denying climate change because of free market values are also the ones that are happiest to endorse over $5 trillion of subsides for the fossil fuel industry per year. It is a myth that there are truly free markets, with many countries happily supporting subsidies and the blocking of imports.

Neoliberalism encapsulates a set of beliefs that include: the need for markets to be free; for state intervention to be as small as possible; strong private property rights; low taxation; and individualism. Underlying neoliberalism is the assumption that it provides a market-based solution that enables everyone to become wealthier. This so-called trickle-down effect has been the central construct of neoliberalism for the past 40 years but there is no evidence that it is happening[3]. Half the world's population lives on less than $5.50 per day. In fact, Oxfam have calculated that the eight richest people in the world currently own the same amount of wealth as 4 billion of the poorest people collectively. The IMF recently declared that the last generation of economic policies may have been a complete failure.

Given the long-term challenges of climate change, biodiversity loss and possible repeat pandemics, the major lesson is that free markets do not protect us. Instead, state intervention guided by experts, incorporating and valuing society and communities, underpinned by supportive and dynamic business is what is required to deal with the climate change and the other challenges of the 21st century. What we need is a new era of planetary stewardship led by Governments and underpinned by new economic theories.

Taking Action

Potential solutions to reduce global carbon emissions were outlined in Chapter 11, but if we are to reach net zero emissions by 2050, we need to implement all of them. Project Downdraw[4] has identified over 80 high-level solutions that can be implemented at a range of scales to achieve the 1050 Gt carbon dioxide emissions removal required by 2050. Figure 58 is based on the work of Avit Bhowmik and colleagues[5], who show how many

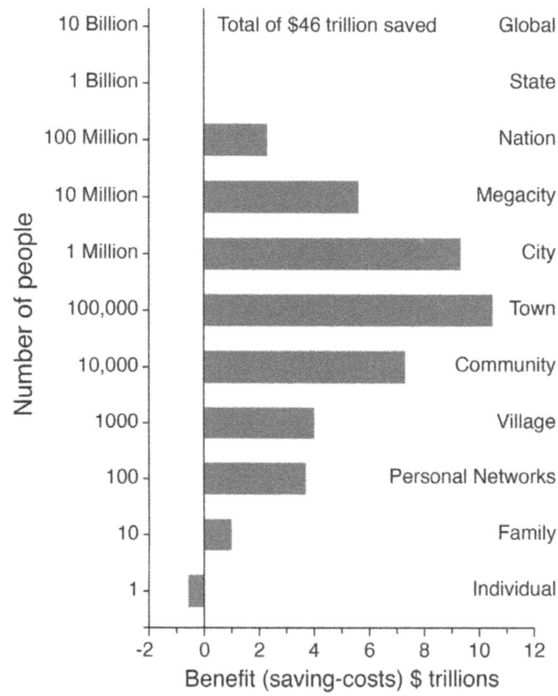

Figure 58. Potential Climate Change Solutions from the Individual to Global Scale

of these can be implemented at each scale — ranging from the individual to the global level. Individual and family actions can remove 14% of the required 1050 Gt CO_2, while action at the town and community scale can remove 31%, and city and state scales can remove 33%. This is a counter-blast to the climate change deniers who suggest that individuals, not companies or Governments, should take on the responsibility for dealing with climate change. This shift in blame allows climate change deniers to continue supporting the fossil fuel industry as it is just meeting a demand in the market. Individual and family actions are important as it shows Governments and corporations that people are serious about climate change, but it is not the solution. Climate change solutions are most effective from the community to the national level. All of these solutions are win-win and taken together, the net benefit, savings minus costs, could be over $46 trillion.

Government, Corporations, and Civil Society

To achieve effective carbon emission reductions requires a partnership between government, both local and national, corporations, and civil society that is supported and encouraged by individual behaviour change. Governments control the aspirations of civil society through the rule of law and the development of policy. It is clear that governments can use incentives, subsidies, taxation, and regulation to make our societies more sustainable and carbon neutral. Governments are also the major driver of innovation, through investment in university research, funding industrial R&D, and driving demand through incentives[2]. Importantly, governments can facilitate the rapid switch from fossil fuels to renewable energy, ensure buildings are carbon neutral, encourage reforestation and rewilding of large areas, promote low-emission farming and a more plant-based diet, and support the very poorest people in society, to help build resilience to climate change impacts.

The world's top 100 companies generate more than $15 trillion in revenue per year. In many ways, businesses control our lives as they influence what we eat, what we buy, what we watch, and even who we vote for. Many are already changing, adopting Science Based Targets[6] so they can achieve net zero carbon emissions by 2050. The challenge for business and industry if they want to remain relevant and trusted in the 21st century is changing their relationship with the environment and society. The classic, linear economic model 'take, make, dispose', relies on large quantities of cheap, easily accessible materials and energy. We are reaching the physical limits of this model. New inclusive economic theories are emerging to show the fundamental issues with the throwaway corporate culture. The circular economy is essential if companies are to be part of the climate change solution. The circular economy minimises the number of resources that are extracted and maximises the value of products and materials throughout their lifecycle, through reuse and recycling. Applying a circular economy could unlock up to €1.8 trillion of value for Europe's economy alone. Companies, therefore, need to plan and make products that have longevity, upgradability, and recyclability built-in. They need to design out waste and pollution.

Though individual actions will only make a small contribution to carbon reductions, they are extremely important as they send a strong message to both government and corporations that citizens want and support major changes. Individual action has had an impact. The School Climate Strikes and the Extinction Rebellion protests have brought together diverse groups of people across the world, all wanting governments to start taking the protecting of our planet seriously. And change is starting to happen, with over 1,400 local governments and over 35 countries having declared that we are in a climate emergency. But we must also remember that not everyone is equally responsible for the current climate crisis. A study by Oxfam showed that 50% of carbon emissions directly related to lifestyle are emitted by the richest 10% of people in the world (Figure 59)[7]. The poorest 50% of our global society emit just 10% of the pollution. Individual action undertaken by the very wealthiest in society could have a major impact on global carbon emissions.

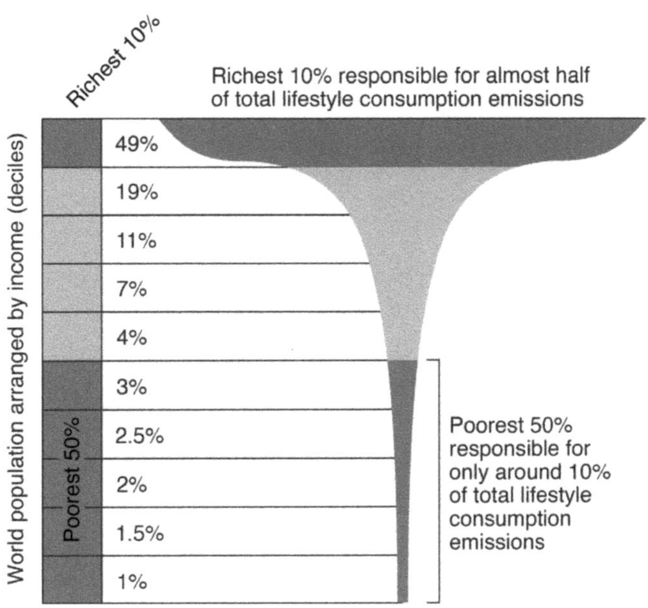

Figure 59. Global Lifestyle Carbon Emissions by Income Group

International Institutions

To support, encourage and when required, enforce positive change, we need international organisations fit for the challenges of the 21st century. Many of these institutions, such as the UN, World Bank, IMF, were formed just after the Second World War[8]. Others such as OECD and OPEC (Organization of the Petroleum Exporting Countries) were formed in the early 1960s[8]. There is a need for these international institutions to represent everyone in the world to ensure fair and equitable governance. The World Bank and IMF could be redesigned so that they focus on developing the green sustainable economy, supporting the net zero targets, and alleviating poverty, with the Sustainable Development Goals at the heart of all their decision-making. The present aim of the World Trade Organization (WTO) is to ensure that trade flows as smoothly, predictably, and freely as possible, but encouraging trade and consumption makes reductions in global carbon emissions harder. It prevents meaningful local, national, and international environmental protections and regulations. This has been challenged recently by President Trump who has imposed tariffs on many countries and sees free trade as counter to the interests of the USA. In light of the recent challenges to the world order, may be the WTO could transition into the World Sustainability Organization (WSO), the first aim of which could be to support and help restructure the economies of countries that rely on fossil fuel exports.

One quick and simple change that could be made is to upgrade the UN Environment Agency. Currently, it has a secondary status within the UN system, below that of trade, health, labour, and even maritime affairs, intellectual property, and tourism. The UN Environment Agency's budget is small, less than a quarter of the UN World Health Organization (WHO)'s budget and a tenth of that of the UN World Food Programme, despite being central to both health and food security. If the UN Environment Agency were to be upgraded to the UN World Environment Organization (WEO), and given a budget at least the size of the WHO's, it could oversee the Sustainability Development Goals, the Convention on Biological Diversity, and the Convention on Climate Change to ensure that they are mutually reinforcing and not in opposition in order to create win-win-win solutions.

Government Solutions

Good governance is our greatest weapon against climate change because Governments control the aspirations of civil society through the rule of law and the development of policy. Incentives, subsidies, taxation, and regulation all play a role in how Governments can make our societies more sustainable and carbon neutral. Governments drive innovation through their support of R&D in Universities, Government laboratories and private companies. Here is a check list of what Governments, both national and local, need to do to reduce GHG emissions while boosting the economy and employment.

1. Support renewable energy
   Governments need to enable the phasing out of fossil fuels, by incentivising and investing in solar, wind, hydro, geo, tidal and other low carbon energy. There are various obvious benefits. First, it reduces air pollution, saves lives and reduces healthcare costs. Around the world over, 10 million people die prematurely due to fossil fuel related air pollution. Second, it increases energy security by reducing dependency on imported fossil fuel. Third, it creates new jobs in dynamic new 21st century industries. Governments also need to reduce overall waste, ensure maximum recycling and full energy recovery from non-recyclable waste. This will prevent waste from going to landfill which can cause water pollution and creates the powerful GHG gas methane. Governments also need to fund research and commercialisation of fusion power. As it is the only real long-term sustainable energy source that can match the huge demands in the 21st century — which by 2050 will double or even triple today's demand.

2. Support electric cars and public transport
   Governments need to invest in electric vehicle infrastructure as this helps the transitions to 100% electric vehicle that is needed to phase out the use of petrol and diesel. This needs to be combined with investment through subsidising public transport and cycling. Again, these actions reduce air pollution and encourage daily exercise, increasing health and wellbeing as well as reducing healthcare costs. Public transport and cycling also can create new jobs and reduce the average commuting time to boost productivity and economic growth.

3. Cut fossil fuel investment and subsidies.
   Governments spend over \$5 trillion per year of taxpayer's money subsiding fossil fuel use. This sum of money could be used to support healthcare, infrastructure projects, renewable energy and other essential Government priorities. Governments need to cut investment in overseas fossil fuel projects and support decarbonisation of other countries by putting overseas investment into sustainable renewable energy. These will stimulate international markets for home-grown renewable energy technology.

4. Tax fossil fuel use
   Governments and economic regions such as the EU need to tax the import and use of oil, coal and natural gas. A carbon tax has been shown by economists to be the most efficient way to cut emissions. This will make renewable energy and electric vehicles even more competitive. At the same time, the increased Government revenue can be channelled into investment in low carbon technology and infrastructure projects. At an international level, agreements should be made to tax Internet and AI companies for their carbon emissions as well as legitimises taxation of international companies while incentivising change in global industries with very large carbon footprint. It would also be sensible to change the Chicago Accord to tax aviation fuel based on carbon emissions per mile. To ensure that it does not skew the global market, set the aviation tax globally and at a very low level to start with. The aviation fuel tax would drive efficiency and the development of new aviation technology, such as the use of biofuels or carbon-neutral synthetic kerosene.

5. Build new low carbon infrastructure
   Governments should invest in new electric high-speed 'bullet' train networks across the world. In many regions and countries, it could replace high carbon flights with low carbon trains. Trains can also provide faster connectivity with reduced security risks. These massive infrastructure projects would create a huge number of jobs. One example that has been given is that 80% of all internal flights in the USA could be replaced by high-speed train networks operating on the East and West Coasts. Governments need to develop new building

regulations to ensure all new buildings are carbon neutral and support low carbon retrofitting of older buildings. This creates healthier homes, offices and factories. It also reduces the energy costs of homes alleviating energy poverty. Finally, Governments need to build new recycling facilities and power station that generate energy from waste plants. Policies are needed to strengthen the circular economy, providing more low carbon energy and preventing waste from going to landfill.

6. Reforest and Rewild
   Policies are required to incentivise replanting of regionally appropriate mixed woodland or forest. This provides natural negative carbon dioxide emissions by storing carbon. It increases and protects local biodiversity, stabilises local soil, reduces the risk of extreme flash floods, increases productivity of adjacent agricultural land and when done on a large scale, stabilises regional rainfall and weather patterns. Governments need to incentivise rewilding of natural habitats such as wetlands, peatlands, grasslands, mangroves and areas of coastal seagrass. All these habitats provide natural storage of carbon and reduce carbon emissions from the land, increase and protect local biodiversity, reduce the risk of extreme storm surges and flooding, reduce the rate of coastal erosion, increase agricultural productivity as well as being aesthetically pleasing and increasing human wellbeing.

   One way which Governments can encourage reforestation and rewilding is by using progressive land taxation to ensure the better use of land in protecting the environment. This financially incentivises reforestation and rewilding, low carbon and sustainable farming as well as active flood management. It can also help to move land ownership back into public ownership to ensure it is used productively for the public good.

7. Promote low emission farming and diet
   Governments must provide a new vision of farming which protects the environment, increases food security, provides healthier diets and increases animal welfare. One approach could be mandatory food labelling that must include the country of origin, carbon footprint, farming method and relative healthiness. This would provide consumers with a real informed choice about the environmental impact and health implications. Another

approach would be to mandate that public procurement must provide only vegetable-based, high nutritional, and environmental certified food. This is something that many organisations, companies and Universities — including my own, have already done. This would incentivise large food companies to produce higher quality vegetable-based food products and increase revenue to environmentally positive farmers. On top of that, it provides healthy diets for people that work for or are cared for by the State (hospitals, old peoples home, schools and prisons). Why not take a step further and also ban advertisement of unhealthy foods and tax unhealthy foods such as meat and sugar? This makes unhealthy foods more expensive and less attractive, thus helping to change the food culture away from unhealthy food. Governments can regulate and financially incentivise farming to restore soils, biodiversity as well as minimise water and air pollution. This reduces GHG emissions from agriculture, increases local biodiversity and helps maintain ecosystem services. It also helps to reduce the impacts of flash flooding and droughts plus reduces the huge cost of water treatment and purification.

By encouraging farming methods that work in harmony with nature such as agroecology, circular agriculture and organic farming, Governments can reduce waste and GHG emissions from agriculture. Governments should tax non-organic pesticides and synthetic fertilisers. We also need a globally ban pesticides list that are harmful to bees and other essential pollinators or food production will be greatly reduced in the future.

8. Support the Emissions Trading Schemes
Around the world, national and regional economic emissions trading or 'cap and trade' have been successful in reducing emissions. These are market-based environmental regulation that allocates permits which companies have to buy if they want to emit GHGs. These permits can be traded to ensure the companies that can reduce their emissions the fastest are financially incentivised to do so. The number of permits can be cut each year and their price is increased to ensure a continual drop in GHG emissions and incentivise rapid innovation. This makes companies more efficient and environmentally friendly, while saving companies and large organisations money, as well as driving innovation and sharing of best practice between organisations.

9. Adaptation — protect your citizens
   As we saw in Chapter 12, this is the forgotten aspect of climate change. It is crucial for Governments to undertake and publish regular national climate change risk assessments. These allow companies, organisations, local Government and individuals to assess the risks of climate change and adapt as appropriate. To make adaption successful, Governments need to inform and educate the public on the enhanced risks due to climate change. They need to help citizens understand the risks of climate change and the actions taken by the Government to protect them. Support is needed for local infrastructure and social adaptations to climate change — these protects the health and wellbeing of citizens as well as the local economy. It is also possible to build hard infrastructure to protect people and property from changing climate such as increased storms, floods, droughts and heatwaves. Other suggestions include building low carbon passive cooled buildings as well as retrofitting air and ground heat exchange air conditioning to existing buildings to protect people from the health effects of heatwaves.

10. Universal Basic Income
    Governments also need new radical suggestions to reduce excess consumption and eliminate extreme poverty. One suggestion is Universal Basic Income (UBI), a guarantee financial payment to every citizen; unconditionally without any obligation to work at a level above their subsistence needs. This has been shown in many case studies to eliminate extreme poverty, breaks the link between work and unsustainable consumption, and allows citizens to retrain anytime in their career — which is essential in the 21st century. It also allows people to choose sustainable and socially responsible jobs, to think long-term, well beyond the next payday, and allows citizens to choose to look after their elderly and sick relatives. UBI also increases creativity and engagement in local communities, reduces social stress, reduces healthcare costs and increases the number of entrepreneurs and their success rates. It is also usually much cheaper than administering heavy judgemental welfare systems that many countries currently have. Of course, Governments need to combine UBI with progressive taxation to encourage people into employment. Additionally, having

free healthcare and education (including technical training and Universities) creates a healthy, sustainable, socially mobile, economically dynamic and entrepreneurial society.

Corporate Positive Power

The world's top 100 companies generate more than $15 trillion in revenue per year. More than half of the world's 100 most valuable companies are located in the USA (35 companies) or China (23 companies). Businesses control our lives: what we eat, what we buy, what we watch and even who we vote for. Corporations have immense power and we should harness this to change the world for the better. Below are suggestions of what your corporation or organisations can do to help save our planet.

1. Ambition and vision
   Short-termism in the corporate world is so last century. Companies need to set the agenda as they are more agile and fast-moving compared to Governments. There is strong evidence that being sustainable and actively caring about the environment is good for business. For example, ten years ago, the Carbon Disclosure Project found that corporations that actively manage and plan for climate change secure a 67% higher return than companies who refuse to disclose their carbon emissions[9]. One great example already given is Microsoft, which has set the agenda for the technology sector. Asian companies are catching up too. Companies that have signed up to the Science Based Targets initiative (SBTi) include Hong Kong real estate firm Swire Properties, Chinese computer giant Lenovo, and Malaysian textile firm Tai Wah Garments Industry. In Singapore, real estate firm City Developments Limited (CDL) and Frasers Property have made a net zero pledge. Even oil and gas giants such as Malaysian Petronas, PetroChina and BP have announced their companies will be net-zero by 2050.

2. Open and transparent business
   Going carbon neutral or carbon negative is a journey that needs to be transparent so employees, clients, customers and regulators can

appreciate what you are doing. Below is a suggested process that you can consider. First, measure your company's carbon footprint. Then, make a realistic plan on how your company can reduce its carbon footprint. Following that, publish both the carbon footprint and the plan before implementing your company's sustainability plan. Finally, publish the company's successes and failures. It is crucial for your company to repeat the cycle until you go carbon negative and then keep going. Companies should apply the same process to other greenhouse gases, water usage and raw material consumption.

3. Set Emission Reduction Targets
   The cornerstone of environmental sustainability is to reduce the GHG emissions of a company's activities in the generation of its products and services. Companies first need to be in compliance with the law and there are numerous regulatory bodies, both at local level, European and international level, that control the reporting of GHG emissions. Companies should set an ambitious long-term emission reduction target that has been approved by the SBTi[6]. Companies should publish an annual environmental audit to the most stringent international standards. Corporations need to ensure this reporting conforms to the recommendations of the Task Force on Climate-related Financial Disclosures (TCFD), the remit of which has now been taken over by the International Financial Reporting Standards (IFRS) Foundation[10]. One approach that has worked for many organisations is to set an internal price on carbon within the company to motivate behavioural change. Companies should initially aim to be a 'carbon neutral' company and progressively increase the targets to become a 'carbon negative' company — meaning the operation of the company results in more carbon being taken out of the atmosphere than is put in — making up for the previous years of emissions.

4. Focus on energy
   Energy consumption is a significant source for the GHG emissions of most companies. It could be from the manufacturing of products, servicing the offices, factories, warehouses and/or datacentres. Green innovation extends way beyond selecting energy rated premises and

companies need to have an energy efficiency strategy. This should cover how to move to renewable energy for electricity through the energy supplier and through power purchase agreements. Moreover, it should also detail a plan on how to develop onsite renewable energy generation where practical as well as how to switch to renewable energy instruments such as Guarantee of Origin (GO) and International Renewable Energy Certificates (I-REC & REC) for electricity. Many consultancy companies advocate switching to Green Gas Certificates for gas but there is considerable doubt if natural gas (methane) can ever be called green; instead, companies should focus on increasingly using renewable energy. For example, companies could get ahead by switching the company's car/van fleet to electric.

5. Apply the circular economy
The linear 'take, make, dispose' economic model relies on large quantities of cheap, easily accessible materials and energy, and is a model that is reaching its physical limits. A circular economy is an attractive and viable alternative that minimises the amount of resources that are extracted and maximises the value of products and materials throughout their lifecycles.
Applying circular economy principles could unlock up to EUR 1.8 trillion of value for Europe's economy. According to The Ellen MacArthur Foundation, business plays a central role in creating the systemic change required to reap the financial benefits of this transition. The key stages for success include: First, plan and make products that have longevity, upgradability and recyclability built-in. Next, design out waste and pollution in your company and products. Finally, make products and materials that can be used for the long-term and are easy to be recycled.

6. Engage employees
CEOs are taking sustainability seriously to ensure recruitment and retainment of the very best employees. Brilliant young people nowadays want to work for a company that has a strong moral and ethical code of conduct. Making companies environmentally sustainable requires profound changes in the company's own behaviour that called

for the creativity of the employees and building a more collaborative mode of operation. Many companies have already established a sound Environment Management System (e.g. ISO14001) to focus on the key aspects and impacts which include: energy use, business travel, waste, and water. Many corporations engage employees through employee ambassadors, green champions, environmental specialists and partner organisations. They listen to their employees, and where practical, embrace and implement their ideas. As the best companies encourage employees to embed environmental sustainability objectives into their daily work and life, many companies also encourage employees to volunteer their free time in promoting the importance of environmental matters in their local community.

7. Link into the supply and value chain
   In the globalised economy, companies rely on their supply chain to meet their business objectives. A company's supply chain GHG emissions could be as much as four times that of the company's (internal) emissions. Environmental innovation and sustainability need to be extended to both the external value and supply chain. This can be done by: First, create a responsible purchasing policy to ensure the purchase of products and services with a low environmental impact. Second, engage with the supply chain to ensure the company is working with other companies that share its sustainable goals. Companies should instigate key performance measures of the supply chain for sustainability and encourage technological developments in the supply chain to help them innovate and produce goods in a greener way. It is also possible now to use blockchain technology to track the environmental impact of the goods and services a company purchases. Every corporation and company should evaluate and audit their supply chain annually against their key performance measures and their carbon emission target.

8. Change the whole conversation
   If companies want to remain relevant and trusted in the 21st century, they need to change their relationship with the environment and society. As discussed in this book and many others, classic economics

is dead, and the throw away corporate culture is also dead. The circular economy is essential if companies are to be a part in saving our planet. Sustainability must become a central tenet of all companies as it increases the longevity and profitability of the company. Each company should have a 5-year, 10-year, 15-year and 50-year plan. The reason why Microsoft[11] are planning to become a carbon neutral company by 2030, and then a carbon negative company by 2050 is simple. They are planning to continue selling us software and operating systems in 2050, and they cannot do that on a dead planet.

9. Influence Governments
   Companies need to make their Government know that they take the future of our planet seriously. Many politicians think short-term, while many companies now think long-term, so we need to ensure new environmental policies are long-term. Companies need to lobby for various sectors to help their Governments think long-term. In the infrastructure sector, companies need to lobby for it to be as sustainable, low carbon and high quality as possible. In terms of a highly skilled workforce to support the massively expanding global green economy, companies can lobby for the Government to have the very best education system possible. In the realm of environmental regulations, companies can lobby for sensible, strict, and long-term regulations which reward companies and organisations that comply fully. Once Governments have set strict environmental standards, companies need to make sure they stick to them so they can plan long-term. Finally, companies need to lobby the Government to ensure independent (non-commercial) enforcement of all environmental regulations so that there is a level playing field to ensure fair competition between companies.

10. Calling all entrepreneurs
   The global green economy is growing at a phenomenal rate and needs new technology and products. Rapid change brings about opportunity and there is no better time than now to be an entrepreneur. We need disruptors, novel thinkers, and new companies to drive green innovation. We need a new wave of social and environmental entrepreneurs to help save our planet.

Power of the Individual

Throughout history, it is individuals that have changed the world. A young girl standing outside the Swedish Parliament protesting about climate change. A black woman refusing to give up a seat for a white man. Individuals are the catalysts that allow the rest of us to demand change. Here are nine suggestions of how you, as an individual can make a difference.

1. Talk about climate change
   The first and most important thing you can do is talk about climate change to everyone. The greatest challenge in the history of our species should not be a taboo subject. We need new solutions, new social structures and new economics to solve this crisis. So, talk about climate change[12].

2. Switch to a more vegetable-based diet.
   Think about your diet for the planet, but also for your own health and longevity. You can become a Flexitarian — having a diet of mostly plant-based foods but allows meat and dairy products in moderation[13]. A western standard meat-based diet produces 7.2 kgCO2e/day, while a vegetarian diet produces 3.8 kgCO2e/day and a vegan diet produces 2.9 kgCO_{2e}/day. In addition, meat production, particularly beef, is a major cause of tropical deforestation. Cutting your meat consumption saves the environment and reduces your carbon emissions. A more vegetable-based diet is also good for you and your family's health because meat, especially highly processed meat, has been linked to: high blood pressure, heart disease, chronic obstructive pulmonary disease, bowel and stomach cancer. If you are not already a vegetarian, aim to become one, but no one will blame you if you occasionally have some meat or fish. Perhaps you can also try dairy-free alternatives as they all have a lower carbon footprint compared to dairy milk.

3. Switch to a renewable energy supplier.
   If it is possible in your country, consider switching to a renewable energy supplier; it may not even cost you anything extra. If we all switched, then energy companies would have to generate more renewable energy to meet demand. You could persuade your workplace, place of worship, local authority, school and sports club to switch to renewable energy.

4. Make your home energy efficient.
   You can save energy (and money) by making your home as heat efficient as possible. First, ensure your home is insulated (roof and walls) to the highest standard. Ensure your windows and doors are fitted properly to avoid draughts. You could then turn the thermostat down a degree and wash at a lower temperature. Other considerations include reducing the use of home appliances, installing a smart meter, and replacing halogen bulbs with more energy-efficient LED bulbs. All of the above will save you money through energy efficiency. If your heating/cooling system at home needs replacing, make sure it is replaced with the most efficient system possible. If finances allow, consider investing in ground and air heat exchangers that can produce either heating or cooling (which you will need as the world heats up).

5. Use cars less
   We should all increase our walking, cycling and use of public transport. This will improve your fitness and your health. If you need a car occasionally, then hire an electric one. If you really need a car, then choose the smallest and most efficient one possible, or if you can afford it, buy an electric car.

6. Reduce flying.
   If possible, choose an alternative form of travel such as the train instead of flying. In big countries such as China or across regions such as Europe, it is easier to take the train and if you take in the time to get to airport and get through security, the train may even be quicker. If you must fly for work, then select only essential trips and consider offsetting the carbon emissions using reputable specialist firms. Alternatively, use the 'offset' money to reduce your emissions at home or work. Many organisations have now decided to offset all unavoidable emissions at ten times (x10) the estimated emissions and to monitor closely their chosen offset schemes. In this way, they can be assured their emissions have been removed.

7. Divest your pension and investments from fossil fuels.
   If you are lucky enough to have a pension or investments, get them to divest from fossil fuels. You can do this by lobbying your pension

fund to divest from all fossil fuel investments. If they refuse, then move your pension fund, as your pension money will be safer and may even earn more. We know the fossil fuel industry will be greatly affected by future climate change legislation and will cease to be profitable. Fossil fuel companies will suffer from what is called 'stranded assets'. The company's nominal worth is based on how much fossil fuels they have in their oil and gas fields as well as their coal mines. These assets will be worthless when fossil fuel mining and drilling are banned due to climate change. Fossil fuels are a poor long-term investment and green companies are already returning twice as much profit as fossil fuel companies.

8. Refuse/reject/reduce/recycle your consumption.
   You do not need all that stuff. Reject the idea that consumption is good for you and that more stuff makes you happier. Think carefully about what you need and how you want to live a sustainable low carbon lifestyle. Competing to have the latest gadget, car or clothes just leads to social stress. Not worrying about what everyone else is buying will reduce your social stress and make you healthier and happier. Friends, family, work and community are what make humans happy. Remember also as a consumer, you have huge power, so exercise that power through your choices. Do you need so much packaging on your goods? Do you need fast fashion? Do you need plastic bottles or disposable coffee cups? Do you need a big SUV car or a truck? Consider buying clothes that are stylish but will last, so you reuse them again and again. Have a water bottle, and a reusable coffee cup. Engage with your local community reuse network as someone local may want to reuse the stuff you no longer need or want. Fix things when they break. Replace your phone battery and/or screen instead of buying a whole new phone. If all of us used our purchasing power only on sustainable brands, all brands would have to become sustainable or go out of business.

9. Protest
   People power is real. The School Climate Strikes and the Extinction Rebellion protests have brought together diverse groups of people

across the world all wanting Governments to start taking the protecting of our planet seriously. Protests are having an impact and change is starting to happen. Demand your politicians to tell you where they stand on dealing with climate change. If you live in a democratic country, you can vote for new Governments. Demand change. Demand action. Use your vote wisely if you are lucky enough to have one.

Conclusion

Climate change is one of the few areas of science that makes us examine the whole basis of modern society. It is a subject that has politicians arguing, sets nations against each other, questions the role of companies in society, queries individual choices of lifestyle, and ultimately asks questions about humanity's relationship with the rest of the planet. Only by working together can we deal with one of the greatest crises that has ever faced humanity. This is a crisis that Governments, Corporations and Civil Society must work together to solve. There is very little doubt that climate change will accelerate in this century; our best estimates suggest a global mean surface temperature rise of between 1.5°C and 5.5°C by the end of the 21st century. Sea level is projected to rise by between 50 cm and 130 cm by 2100, with significant changes in weather patterns, and more extreme climate events. World leaders have pledged to keep climate change to less than 2°C and if at all possible, below 1.5°C. This book has demonstrated that we have the science to understand the causes, consequences and potential solutions for climate change. We have the technology, the resources, and the money to deal with climate change. What we currently lack are the political will and policies to enable all the positive win-win solutions needed to make a better, safer, healthier and hopefully, happier world. With the growing awareness of the environmental crisis facing the planet, the public pressure for change is growing and new policies and ways of thinking are starting to emerge. The question is whether they will be quick enough to get the world to net zero carbon emissions by 2050.

References

Chapter 1. What is Climate?

1. T. R. Oke, *Boundary Layer Climates*, Routledge, p. 464 (2001).
2. https://www.gov.uk/government/publications/heat-mortality-monitoring-reports/heat-mortality-monitoring-report-2022
3. R. G. Barry and R. J. Chorley, *Atmosphere, Weather and Climate*, Routledge, p. 536 (2009).
4. N. Wells, *The Atmosphere and Ocean: A Physical Introduction* Wiley, p. 432 (2011).
5. https://www.noaa.gov/jetstream/atmosphere
6. https://www.noaa.gov/jetstream/global/jet-stream
7. https://ugc.berkeley.edu/background-content/ocean-circulation/

Chapter 2. What is Climate Change?

1. IPCC, Climate Change 2021 — The Physical Science Basis Contribution of Working Group I to the Sixth Assessment Report of the Intergovernmental Panel on Climate Change. https://www.ipcc.ch/report/ar6/wg1/
2. A. Bell, *Our biggest experiment: A history of the climate crisis*, Bloomsbury, p. 384 (2022).
3. Weart, S. R. *The Discovery of Global Warming, New Histories of Science, Technology, and Medicine*, Harvard University Press (2008).
4. https://news.climate.columbia.edu/2018/06/26/james-hansens-climate-warning-30-years-later/

Chapter 3. Weather Versus Climate

1. J. Gleick, *Chaos: Making a New Science*, Penguin, p. 384 (2008).
2. https://www.snexplores.org/article/explainer-chaos-theory-math-physics-nature
3. https://www.ncei.noaa.gov/access/monitoring/nao/
4. https://www.ncei.noaa.gov/access/monitoring/pdo/
5. https://www.ncei.noaa.gov/access/monitoring/enso/
6. https://nhess.copernicus.org/articles/19/715/2019/
7. https://voxdev.org/topic/energy-environment/climate-adaptation-bangladesh

Chapter 4. Past Climate Changes

1. C. H. Langmuir and W. Broecker, *How to Build a Habitable Planet: The Story of Earth from the Big Bang to Humankind*, Princeton University Press, revised and expanded edition, p. 720 (2012).
2. T. Lenton, *Earth System Science: A Very Short Introduction*, OUP (2016).
3. J. J. Lowe and M. Walker, *Reconstructing Quaternary Environments*, Prentice Hall, p. 472 (1997).
4. W. F. Ruddiman, *Earth's Climate: Past and Future*, W. H. Freeman, p. 464 (2017).
5. M. Maslin 'Chapter 3: How the ice age began', In *The Complete Ice Age: How climate change shaped the world* (edited by Brain Fagan), Thames and Hudson, 48–61 (2009).
6. M. Maslin 'Chapter 4: The climatic rollercoaster', In *The Complete Ice Age: How climate change shaped the world* (edited by Brain Fagan), Thames and Hudson, 62–91 (2009).
7. Ruddiman, W.F. *Plows, Plagues, and Petroleum: How Humans Took Control of Climate*, Princeton Science Library (2016).
8. Lewis, S.L. and M.A. Maslin The Human Planet: How humans caused the Anthropocene (Penguin and Yale University Press, 2018).
9. P. Gibbard, et al., The Anthropocene is a Geological Event, not an Epoch, *Episodes* https://doi.org/10.18814/epiiugs/2021/021029 (2021).

Chapter 5. Evidence for Climate Change

1. IPCC, Climate Change 2021 — The Physical Science Basis Contribution of Working Group I to the Sixth Assessment Report of the Intergovernmental Panel on Climate Change. https://www.ipcc.ch/report/ar6/wg1/

2. Mark Maslin, John Lang and Fiona Harvey. A short history of the successes and failures of the international climate change negotiations. *UCL Open: Environment* (2023) https://journals.uclpress.co.uk/ucloe/article/id/904/
3. Gupta, J. *The History of Global Climate Governance* (Cambridge University Press, 2014).
4. Emanuel, K. *What We Know about Climate Change: updated edition* (The MIT Press, 2018).
5. Ripple, W., et al. The 2023 state of the climate report: Entering uncharted territory, BioScience, 2023, biad080, https://doi.org/10.1093/biosci/biad080
6. https://berkeleyearth.org/global-temperature-report-for-2024/
7. https://www.nature.com/articles/s41586-019-1401-2
8. https://www.metoffice.gov.uk/about-us/news-and-media/media-centre/weather-and-climate-news/2022/kyoto-cherry-blossom-dates-shifted-by-human-influence
9. https://interactive.carbonbrief.org/attribution-studies/index.html
10. https://www.pnas.org/doi/10.1073/pnas.1920849117
11. https://en.wikipedia.org/wiki/Typhoon_Haiyan
12. https://www.climate.gov/news-features/understanding-climate/climate-change-global-temperature

Chapter 6. Modelling Future Climate

1. Gettelman A. and R. B. Rood. Demystifying Climate Models: A Users Guide to Earth System Models. Springer Open (2016).
2. IPCC, Climate Change 2021 — The Physical Science Basis Contribution of Working Group I to the Sixth Assessment Report of the Intergovernmental Panel on Climate Change.
3. Lenton, T. *Earth System Science: A Very Short Introduction* (OUP, 2016).
4. A. E. Dessler, *The Science and Politics of Global Climate Change: A Guide to the Debate*, Cambridge University Press (2019).
5. McGuffie K. and A. Henderson-Sellers, The Climate Modelling Primer 4th edition. Wiley-Blackwell, p. 464 (2014).
6. https://www.climate.gov/news-features/climate-qa/couldnt-sun-be-cause-global-warming
7. https://www.sciencedirect.com/science/article/pii/S0959378016300681
8. Stainforth, D., Predicting Our Climate Future: What We Know, What We Don't Know, And What We Can't Know. OUP, p. 368 (2023).

Chapter 7. Climate Change Impacts

1. IPCC, Climate Change 2021 — Impacts, Adaptation, and Vulnerability, Contribution of Working Group II to the Sixth Assessment Report of the Intergovernmental Panel on Climate Change.
2. https://www.nccs.gov.sg/singapores-climate-action/impact-of-climate-change-in-singapore/
3. https://www.climate.go.kr/home/cc_data/2020/Korean_Climate_Change_Assessment_Report_2020_2_eng_summary.pdf
4. https://unfccc.int/sites/default/files/resource/China_NC4_English.pdf
5. https://toolkit.climate.gov/NCA5
6. https://www.ukcip.org.uk/
7. Bell A., Our biggest experiment: A history of the climate crisis, Bloomsbury, p. 384 (2022).
8. Romanello, M., et al. The 2025 Report of the Lancet Countdown on Health and Climate Change. *The Lancet* (2025).
9. Costello, A., et al. Managing the Health Effects of Climate Change, *The Lancet*, 373 (2009): 1693–733.
10. https://www.oecd.org/en/publications/2025/06/global-drought-outlook_28488e98/full-report/towards-a-drier-world_6f10e2e5.html
11. Emanuel, K. *What We Know about Climate Change: updated edition* (The MIT Press, 2018).
12. IPCC, 2019: IPCC Special Report on the Ocean and Cryosphere in a Changing Climate Cambridge University Press, Cambridge, UK and New York, NY, USA, 755 pp. https://doi.org/10.1017/9781009157964.
13. IPCC, 2019: Climate Change and Land: an IPCC special report on climate change, desertification, land degradation, sustainable land management, food security, and greenhouse gas fluxes in terrestrial ecosystems. Cambridge University Press, Cambridge, UK and New York, NY, USA, 755 pp.
14. IPCC, Climate Change 2021 — The Physical Science Basis Contribution of Working Group I to the Sixth Assessment Report of the Intergovernmental Panel on Climate Change.
15. Garcia R.A., et al. Multiple Dimensions of Climate Change and their Implications for Biodiversity, *Science*, 344 (2014): 486–96.
16. https://www.nature.com/articles/nature02121
17. https://www.sciencedirect.com/science/article/pii/S2514664525000190
18. Clarke, B., Otto, F., Stuart-Smith, R. and Harrington, L., 2022. Extreme weather impacts of climate change: an attribution perspective. Environmental Research: Climate, 1(1), p. 012001.

Chapter 8. Climate Tipping Points

1. Lenton, T. *Earth System Science: A Very Short Introduction* (OUP, 2016).
2. Armstrong McKay, D.I. et al. Exceeding 1.5°C global warming could trigger multiple climate tipping points. Science 377, 7950 (2022) DOI:10.1126/science.abn7950
3. IPCC, 2019: IPCC Special Report on the Ocean and Cryosphere in a Changing Climate Cambridge University Press, Cambridge, UK and New York, NY, USA, 755 pp. https://doi.org/10.1017/9781009157964.
4. Maslin, M., et al., Gas hydrates: Past and Future Geohazard? *Phil. Transactions A of the Royal Society*, 368, 2369–2393 doi: 10.1098/rsta.2010.0065 (2010).
5. C. H. Langmuir and W. Broecker, *How to Build a Habitable Planet: The Story of Earth from the Big Bang to Humankind*, Princeton University Press, revised and expanded edition, p. 720 (2012).
6. https://forestplots.net/
7. https://royalsocietypublishing.org/doi/full/10.1098/rstb.2008.0011
8. https://www.carbonbrief.org/significant-risk-of-amazon-forest-dieback-if-global-warming-overshoots-1-5c/
9. https://www.carbonbrief.org/last-refuges-for-coral-reefs-to-disappear-above-1-5c-of-global-warming-study-finds/
10. OECD, Climate Tipping Points: insights for effective policy action. OECD Publishing, Paris, https://doi.org/10.1787/abc5a69e-en (2022).

Chapter 9. Politics of Climate Change

1. IPCC, Climate Change 2022 — Mitigation of Climate Change, Contribution of Working Group III to the Sixth Assessment Report of the Intergovernmental Panel on Climate Change.
2. Bell A., Our biggest experiment: A history of the climate crisis, Bloomsbury, p. 384 (2022).
3. https://www.margaretthatcher.org/document%2F107817
4. Mark Maslin, John Lang and Fiona Harvey. A short history of the successes and failures of the international climate change negotiations. *UCL Open: Environment* (2023) https://journals.uclpress.co.uk/ucloe/article/id/904/
5. Stern, N. *The Economics of Climate Change: The Stern Review*, Cambridge University Press (2007).
6. https://drawdown.org/

7. https://timjackson.org.uk/ecological-economics/pwg/
8. Raworth, K. *Doughnut Economics: Seven Ways to Think Like a 21st-Century Economist*, Random House, p. 384 (2018).
9. Oreskes, N. and M. Conway *Merchants of Doubt: How a Handful of Scientists Obscured the Truth on Issues from Tobacco Smoke to Global Warming* (Bloomsbury, 2012).
10. Corfee-Morlot, J., M.A. Maslin, J. Burgess "Climate Science in the Public Sphere", *Philosophical Transactions A of the Royal Society* 365 no. 1860, 2741–2776. (2007) doi:10.1098/rsta.2007.2084
11. Mann, M. *The New Climate War: The Fight to Take Back Our Planet* (PublicAffairs, 2021).
12. https://www.undp.org/publications/peoples-climate-vote-2024
13. Figueres, C. and T. Rivett-Carnac. *The Future We Choose: Surviving the Climate Crisis* (Manilla Press, 2020).
14. IPCC, 2018. Global Warming of 1.5°C Cambridge University Press, https://doi.org/10.1017/9781009157940
15. IPCC, 2019: Climate Change and Land: an IPCC special report on climate change, desertification, land degradation, sustainable land management, food security, and greenhouse gas fluxes in terrestrial ecosystems. Cambridge University Press, Cambridge, UK and New York, NY, USA, 755 pp.
16. IPCC, 2019: IPCC Special Report on the Ocean and Cryosphere in a Changing Climate Cambridge University Press, Cambridge, UK and New York, NY, USA, 755 pp. https://doi.org/10.1017/9781009157964.

Chapter 10. International Climate Negotiations

1. Mark Maslin, John Lang and Fiona Harvey. A short history of the successes and failures of the international climate change negotiations. *UCL Open: Environment* (2023) https://journals.uclpress.co.uk/ucloe/article/id/904/
2. Vanhala, L. *Governing the End: The Making of Climate Change Loss and Damage* (Chicago Series in Law and Society), University of Chicago Press, p. 251 (2025).
3. https://www.margaretthatcher.org/document%2F107817
4. Gupta, J. *The History of Global Climate Governance* (Cambridge University Press, 2014).
5. Meyer, A. *Contraction and Convergence: The Global Solution to Climate Change* (Green Books, 2015).

6. Figueres, C. and T. Rivett-Carnac. *The Future We Choose: Surviving the Climate Crisis* (Manilla Press, 2020).
7. https://theconversation.com/five-things-you-need-to-know-about-the-glasgow-climate-pact-171799
8. IPCC, Climate Change 2021 — Impacts, Adaptation, and Vulnerability, Contribution of Working Group II to the Sixth Assessment Report of the Intergovernmental Panel on Climate Change.
9. IPCC, Climate Change 2022 — Mitigation of Climate Change, Contribution of Working Group III to the Sixth Assessment Report of the Intergovernmental Panel on Climate Change.
10. https://theconversation.com/cop27-will-be-remembered-as-a-failure-heres-what-went-wrong-194982
11. https://theconversation.com/cop28-five-major-outcomes-from-the-latest-un-climate-summit-219655
12. https://theconversation.com/cop29-five-critical-issues-still-left-hanging-after-an-underwhelming-un-climate-summit-in-azerbaijan-244350
13. Giddens, A. *The Politics of Climate Change*, 2nd edn (Polity Press, 2011).

Chapter 11. Solutions: Mitigation

1. Grubb, M. *Planetary Economics: Energy, Climate Change and the Three Domains of Sustainable Development* (Routledge, 2014).
2. https://www.iea.org/reports/world-energy-outlook-2020
3. Hawken, P., *Drawdown: The Most Comprehensive Plan Ever Proposed to Reverse Global Warming* (Penguin, 2018).
4. Gates, B. How to Avoid a Climate Disaster: The Solutions We Have and the Breakthroughs We Need, Penguin, p. 272 (2022).
5. https://www.pnas.org/doi/suppl/10.1073/pnas.1208993109
6. IPCC, Climate Change 2022 — Mitigation of Climate Change, Contribution of Working Group III to the Sixth Assessment Report of the Intergovernmental Panel on Climate Change.
7. C. Goodall, *What We Need to Do Now: For a Zero Carbon Future* (Profile Books, 2020).
8. https://www.thetransportpolitic.com/2009/07/27/new-u-s-high-speed-rail-association-presents-network-plan/
9. D. Helm, *Net Zero: How We Stop Causing Climate Change* (William Collins, 2020).

10. https://www.imf.org/en/Topics/climate-change/energy-subsidies
11. https://www.ccag.earth/reports/voluntary-carbon-markets-pitfalls-potential-and-the-path-forward
12. https://www.science.org/doi/10.1126/science.aax0848
13. https://www.science.org/doi/10.1126/science.aaz0388
14. Doerr, J., Speed and Scale: A Global Action Plan for Solving Our Climate Crisis Now, Penguin Business, p. 448 (2021).

Chapter 12. Solutions: Adaptation

1. IPCC, Climate Change 2021 — Impacts, Adaptation, and Vulnerability, Contribution of Working Group II to the Sixth Assessment Report of the Intergovernmental Panel on Climate Change.
2. Fisher S., Sink or Swim: How the world needs to adapt to a changing climate. Bloomsbury Sigma, p. 288, 2025.
3. Roaf, S., et al. *Adapting Building and Cities for Climate Change* (Routledge, 2009).
4. Figueres, C. and T. Rivett-Carnac. *The Future We Choose: Surviving the Climate Crisis* (Manilla Press, 2020).
5. Costello, A., et al. Managing the Health Effects of Climate Change, *The Lancet*, 373 (2009): 1693–733.
6. Cripps, E. What does climate justice mean and why should we care. Bloomsbury Continuum, p. 224 (2022).
7. https://theconversation.com/stabilising-the-global-population-is-not-a-solution-to-the-climate-emergency-but-we-should-do-it-anyway-126446
8. https://www.un.org/en/global-issues/population
9. https://www.turing.ac.uk/news/publications/climate-aware-and-resilient-national-security-challenges-21st-century
10. https://www.unep.org/resources/adaptation-gap-report-2024
11. Royal Society, People and the Planet, The Royal Society Science Policy Centre Report 01/12 (2012): 81.

Chapter 13. Solutions: Geoengineering

1. Royal Society, Geoengineering the Climate: Science, Governance and Uncertainty, The Royal Society Science Policy Centre Report, The Royal Society, 10/09 (2009): 81.

2. Morton, O. *The Planet Remade: How Geoengineering Could Change the World* (Granta, 2016).
3. https://www.science.org/doi/10.1126/science.236.4798.169
4. https://www.whoi.edu/oceanus/feature/fertilizing-the-ocean-with-iron/ H.J. Buck, *After Geoengineering: Climate Tragedy, Repair, and Restoration* (Verso 2019).
5. https://www.whoi.edu/oceanus/feature/fertilizing-the-ocean-with-iron/
6. https://climeworks.com/direct-air-capture
7. https://www.sciencenews.org/article/swarm-umbrellas-vs-global-warming-astronomer-thinks-small-save-earth
8. https://www.ciel.org/govern-geoengineering/
9. https://wires.onlinelibrary.wiley.com/doi/10.1002/wcc.707

Chapter 14. Changing Our Future

1. https://timjackson.org.uk/ecological-economics/pwg/
2. Mazzucato, M. *The Value of Everything: Making and Taking in the Global Economy* (Penguin 2019).
3. https://www.jasonhickel.org/the-divide
4. https://drawdown.org/
5. https://iopscience.iop.org/article/10.1088/1748-9326/ab9ed0
6. https://sciencebasedtargets.org/
7. https://www.oxfamnovib.nl/Redactie/Pdf/mb-extreme-carbon-inequality-021215-embargo-en.pdf
8. Lewis, S.L. and M.A. Maslin The Human Planet: How humans caused the Anthropocene (Penguin and Yale University Press, 2018).
9. https://www.theguardian.com/sustainable-business/2014/sep/23/business-companies-profit-cdp-report-climate-change-sustainability
10. https://www.ifrs.org/sustainability/tcfd/
11. https://blogs.microsoft.com/blog/2020/01/16/microsoft-will-be-carbon-negative-by-2030/
12. Katharine Hayhoe, Saving Us: A Climate Scientist's Case for Hope and Healing in a Divided World, Atria/One Signal Publishers, p. 320 (2021).
13. https://theconversation.com/which-diet-will-help-save-our-planet-climatarian-flexitarian-vegetarian-or-vegan-186772

Further Reading

What is Climate?

R. G. Barry and R. J. Chorley, *Atmosphere, Weather and Climate*, Routledge, 9th edition, p. 536 (2009).

T. R. Oke, *Boundary Layer Climates*, Routledge, second (reprinted) edition, p. 464 (2001).

N. Wells, *The Atmosphere and Ocean: A Physical Introduction* Wiley, 3rd Edition, p. 432 (2011).

What is Climate Change?

A. Bell, Our biggest experiment: A history of the climate crisis, Bloomsbury, p. 384 (2022).

Corfee-Morlot, J., et al. Climate Science in the Public Sphere, *Philosophical Transactions A of the Royal Society*, 365/1860: 2741–76 (2007).

A. E. Dessler, *The Science and Politics of Global Climate Change: A Guide to the Debate*, 3rd edition (Cambridge University Press, 2019).

IPCC, Climate Change 2021 — The Physical Science Basis Contribution of Working Group I to the Sixth Assessment Report of the Intergovernmental Panel on Climate Change.

Maslin, M. A. and S. Randalls (eds) *Routledge Major Work Collection. Future Climate Change: Critical Concepts in the Environment* (4 Volumes containing reproductions of 85 of the most important papers published on climate change) (Routledge, 2012).

Weart, S. R. *The Discovery of Global Warming, New Histories of Science, Technology, and Medicine* (Harvard University Press, 2008).

Weather Versus Climate

S. Burt, *The weather observer's handbook*, 2nd edition, Cambridge University Press, p. 506 (2024).

J. D. Cox, *Weather for Dummies*, second edition, p. 400 (2021).

R. Hamblyn, *The Pocket Cloud Book Updated Edition: How to Understand the Skies in association with the Met Office*, David & Charles, p. 144 (2023).

Past Climate Changes

C. H. Langmuir and W. Broecker, *How to Build a Habitable Planet: The Story of Earth from the Big Bang to Humankind*, Princeton University Press, revised and expanded edition, p. 720 (2012).

T. Lenton, *Earth System Science: A Very Short Introduction* (OUP, 2016).

J. J. Lowe and M. Walker, *Reconstructing Quaternary Environments*, Prentice Hall, 2nd edition, p. 472 (1997).

Lewis, S.L. and M.A. Maslin The Human Planet: How humans caused the Anthropocene (Penguin and Yale University Press, 2018).

Ruddiman, W.F. *Plows, Plagues, and Petroleum: How Humans Took Control of Climate* (Princeton Science Library, 2016).

W. F. Ruddiman, *Earth's Climate: Past and Future*, W. H. Freeman, 3rd edition, p. 464 (2017).

Evidence for Climate Change

Archer, D. *Global Warming: Understanding the Forecast*, 2nd edn (John Wiley & Sons, 2011).

A. E. Dessler, *The Science and Politics of Global Climate Change: A Guide to the Debate*, 3rd edition (Cambridge University Press, 2019).

Emanuel, K. *What We Know about Climate Change: updated edition* (The MIT Press, 2018).

IPCC, Climate Change 2021 — Impacts, Adaptation, and Vulnerability, Contribution of Working Group II to the Sixth Assessment Report of the Intergovernmental Panel on Climate Change.

Ritchie, H., Not the end of the World. Vintage, p. 352 (2024).

N. Stern, *The Economics of Climate Change: The Stern Review*, Cambridge University Press, p. 692 (2007).

Modelling Future Climate

Gettelman A. and R. B. Rood. Demystifying Climate Models: A Users Guide to Earth System Models. Springer Open (2016).

IPCC, Climate Change 2021 — The Physical Science Basis Contribution of Working Group I to the Sixth Assessment Report of the Intergovernmental Panel on Climate Change.

Lenton, T. *Earth System Science: A Very Short Introduction* (OUP, 2016).

McGuffie K. and A. Henderson-Sellers, The Climate Modelling Primer 4th edition. Wiley-Blackwell, p. 464 (2014).

Stainforth, D., Predicting Our Climate Future: What We Know, What We Don't Know, And What We Can't Know. OUP, p. 368 (2023).

Climate Change Impacts

Costello, A., et al. Managing the Health Effects of Climate Change, *The Lancet*, 373 (2009): 1693–733.

Garcia R.A., et al. Multiple Dimensions of Climate Change and their Implications for Biodiversity, *Science*, 344 (2014): 486–96.

Clarke, B., Otto, F., Stuart-Smith, R. and Harrington, L., 2022. Extreme weather impacts of climate change: an attribution perspective. Environmental Research: Climate, 1(1), p. 012001.

IPCC, 2018. Global Warming of 1.5°C Cambridge University Press, https://doi.org/10.1017/9781009157940

IPCC, 2019: IPCC Special Report on the Ocean and Cryosphere in a Changing Climate Cambridge University Press, Cambridge, UK and New York, NY, USA, 755 pp. https://doi.org/10.1017/9781009157964.

IPCC, 2019: Climate Change and Land: an IPCC special report on climate change, desertification, land degradation, sustainable land management, food security, and greenhouse gas fluxes in terrestrial ecosystems. Cambridge University Press, Cambridge, UK and New York, NY, USA, 755 pp.

IPCC, Climate Change 2021 — Impacts, Adaptation, and Vulnerability, Contribution of Working Group II to the Sixth Assessment Report of the Intergovernmental Panel on Climate Change.

Romanello, M., et al. The 2025 Report of the Lancet Countdown on Health and Climate Change. *The Lancet* (2025).

Stern, N. *The Economics of Climate Change: The Stern Review* (Cambridge University Press, 2007).

Climate Tipping Points

Armstrong McKay, D.I. et al. Exceeding 1.5°C global warming could trigger multiple climate tipping points. Science 377, eabn7950(2022). DOI:10.1126/science.abn7950

Lenton, T. *Earth System Science: A Very Short Introduction* (OUP, 2016).

OECD, Climate Tipping Points: insights for effective policy action. OECD Publishing, Paris, https://doi.org/10.1787/abc5a69e-en (2022).

Ripple, W., et al. The 2023 state of the climate report: Entering uncharted territory, BioScience, 2023, biad080, https://doi.org/10.1093/biosci/biad080

Politics of Climate Change

C. Figueres, and T. Rivett-Carnac. *The Future We Choose: Surviving the Climate Crisis* (Manilla Press, 2020).

Giddens, A. *The Politics of Climate Change*, 2nd edn (Polity Press, 2011).

Grubb, M. *Planetary Economics: Energy, Climate Change and the Three Domains of Sustainable Development* (Routledge, 2014).

IPCC, Climate Change 2022 — Mitigation of Climate Change, Contribution of Working Group III to the Sixth Assessment Report of the Intergovernmental Panel on Climate Change.

Leggett, J. K. *The Winning of the Carbon War: Power and Politics on the Front Lines of Climate and Clean Energy* (Crux Publishing, 2018).

Klein, N. *On Fire: The Burning Case for a Green New Deal* (Allen Lane, 2019).

Metcalf, G.E. *Paying for Pollution: Why a Carbon Tax is Good for America* (OUP, 2019).

Mann, M. *The New Climate War: The Fight to Take Back Our Planet* (PublicAffairs, 2021).

Oreskes, N. and M. Conway *Merchants of Doubt: How a Handful of Scientists Obscured the Truth on Issues from Tobacco Smoke to Global Warming* (Bloomsbury, 2012).

International Climate Negotiations

Gupta, J. *The History of Global Climate Governance* (Cambridge University Press, 2014).

Meyer, A. *Contraction and Convergence: The Global Solution to Climate Change* (Green Books, 2015).

Vanhala, L. *Governing the End: The Making of Climate Change Loss and Damage* (Chicago Series in Law and Society), University of Chicago Press, p. 251 (2025).

Solutions: Mitigation

Doerr, J., Speed and Scale: A Global Action Plan for Solving Our Climate Crisis Now, Penguin Business, p. 448 (2021).

Gates, B. How to Avoid a Climate Disaster: The Solutions We Have and the Breakthroughs We Need, Penguin p. 272 (2022).

C. Goodall, *What We Need to Do Now: For a Zero Carbon Future* (Profile Books, 2020).

P. Hawken, *Drawdown: The Most Comprehensive Plan Ever Proposed to Reverse Global Warming* (Penguin, 2018).

D. Helm, *Net Zero: How We Stop Causing Climate Change* (William Collins, 2020).

IPCC, Climate Change 2022 — Mitigation of Climate Change, Contribution of Working Group III to the Sixth Assessment Report of the Intergovernmental Panel on Climate Change.

M. Grubb, *Planetary Economics: Energy, Climate Change and the Three Domains of Sustainable Development* (Routledge, 2014).

Solutions: Adaptation

Cripps, E. What does climate justice mean and why should we care. Bloomsbury Continuum, p. 224 (2022).

Doerr, J., Speed and Scale: A Global Action Plan for Solving Our Climate Crisis Now, Penguin Business, p. 448 (2021).

Fisher S., Sink or Swim: How the world needs to adapt to a changing climate. Bloomsbury Sigma, p. 288 (2025).

IPCC, Climate Change 2021 — Impacts, Adaptation, and Vulnerability, Contribution of Working Group II to the Sixth Assessment Report of the Intergovernmental Panel on Climate Change.

Roaf, S., et al. *Adapting Building and Cities for Climate Change* (Routledge, 2009).

Royal Society, People and the Planet, The Royal Society Science Policy Centre Report 01/12 (2012): 81.

Solutions: Geoengineering

H.J. Buck, *After Geoengineering: Climate Tragedy, Repair, and Restoration* (Verso 2019).

Centre for Alternative Technology, *Zero Carbon Britain: Rising to the Climate Emergency* (CAT Publications, 2019). https://www.cat.org.uk/new-report-zero-carbon-britain-rising-to-the-climate-emergency/

Morton, O. *The Planet Remade: How Geoengineering Could Change the World* (Granta, 2016).

Royal Society, Geoengineering the Climate: Science, Governance and Uncertainty, The Royal Society Science Policy Centre Report, The Royal Society, 10/09 (2009): 81.

Changing Our Future

M. Berners-Lee, *There Is No Planet B: A Handbook for the Make or Break Years* (CUP, 2021).

L. Cole, Who cares wins: Reason for optimism in our changing world (Penguin, 2020).

Katharine Hayhoe, Saving Us: A Climate Scientist's Case for Hope and Healing in a Divided World, Atria/One Signal Publishers, p. 320 (2021).

Kennedy-Woodard M. and Kennedy-Williams, Turning the tide on climate anxiety.

M. Lynas, *Our Final Warning: Six Degrees of Climate Emergency* (Fourth Estate, 2020).

Mazzucato, M. *The Value of Everything: Making and Taking in the Global Economy* (Penguin 2019).

Raworth, K. *Doughnut Economics: Seven Ways to Think Like a 21st-Century Economist*, Random House, p. 384 (2018).

Index

www.ingramcontent.com/pod-product-compliance
Lightning Source LLC
LaVergne TN
LVHW051056210926
848486LV00002B/126